The Impact of Critical Rationalism

Raphael Sassower • Nathaniel Laor
Editors

The Impact of Critical Rationalism

Expanding the Popperian Legacy through the Works of Ian C. Jarvie

palgrave
macmillan

Editors
Raphael Sassower
University of Colorado
Colorado Springs, CO, USA

Nathaniel Laor
Tel Aviv University
Tel Aviv, Israel

ISBN 978-3-030-08108-9 ISBN 978-3-319-90826-7 (eBook)
https://doi.org/10.1007/978-3-319-90826-7

Softcover re-print of the Hardcover 1st edition 2019

Cover designed by Thomas Howey

Printed on acid-free paper

This Palgrave Macmillan imprint is published by the registered company Springer International Publishing AG part of Springer Nature.
The registered company address is: Gewerbestrasse 11, 6330 Cham, Switzerland

Contents

List of Contributors

Joseph Agassi Department of Philosophy, Tel Aviv University, Tel Aviv, Israel and York University, Toronto, ON, Canada

Nimrod Bar-Am Rhetoric and Philosophy of Communication Unit, Communication Department of Sapir Academic College, Israel

Jeanette Bicknell Independent Scholar, Toronto, ON, Canada

Lawrence Boland Department of Economics, Simon Fraser University, Burnaby, BC, Canada

Augustine Brannigan Department of Sociology, University of Calgary, Calgary, AB, Canada

Mario Bunge Department of Philosophy, McGill University, Montreal, QC, Canada

Rafe Champion Independent Scholar, Sydney, NSW, Australia

José A. Colen Fundação para a Ciência e a Tecnologia, Centre for Ethics, Politics and Society, Minho University, Braga, Portugal and ICS – University of Navarra, Navarra, Spain

Fred D'Agostino Institute for Advanced Studies in the Humanities, University of Queensland, Brisbane, QLD, Australia

Steve Fuller Auguste Comte Chair in Social Epistemology at the Department of Sociology, University of Warwick, Coventry, UK

Malachi Haim Hacohen Department of History, Political Science, and Religion, The Kenan Institute, Duke University, Durham, NC, USA

John A. Hall James McGill Professor of Sociology, McGill University, Montreal, QC, Canada

Jagdish Hattiangadi Department of Philosophy, York University, Toronto, ON, Canada

Byron Kaldis Department of Philosophy, The National Technical University of Athens, Athens, Greece and Department of Philosophy, Moral Culture Research Institute, Hunan Normal University, Changsha, Hunan Province, China

Nathaniel Laor Departments of Medical Education, Psychiatry, and Philosophy, Tel Aviv University, Tel Aviv, Israel and Child Study Center, Yale University, New Haven, CT, USA

Paul Levinson Department of Communication & Media Studies, Fordham University, Bronx, NY, USA

Scott Nelson Independent Scholar, Vienna, Austria

Zuzana Parusniková Institute of Philosophy, The Czech Academy of Sciences, Prague, Czech Republic

Sheldon Richmond Independent Scholar, Thornhill, ON, Canada

Paul A. Roth Department of Philosophy and Jewish Studies, University of California, Santa Cruz, CA, USA

Raphael Sassower Department of Philosophy, University of Colorado, Colorado Springs, CO, USA

Jeremy Shearmur School of Philosophy, Australian National University, Canberra, ACT, Australia

Stephen Turner Department of Philosophy, University of South Florida, Tampa, FL, USA

John Wettersten School of Social Sciences, University of Mannheim, Mannheim, Germany

H. T. Wilson Professor of Public Policy & Public Law, Program Evaluation & Policy Analysis, York University, Toronto, ON, Canada

Kei Yoshida Faculty of Social Sciences, Waseda University, Tokyo, Japan

Jesús Zamora Bonilla Faculty of Philosophy, UNED, National Distance Education University, Madrid, Spain

CHAPTER 1

Introduction: The Legacy of Ian C. Jarvie

Nathaniel Laor and Raphael Sassower

Ian C. Jarvie has suggested that "All that there is the flux of experience, bewildering and incoherent as it presents itself to us, and there is our yearning to give it order and thus to be able to enjoy it, to no longer be afraid of and threatened by it" (Jarvie 1981, p. 234). "I take it that all human encounter is self-making, and human encounter with the ideas and attitudes of others can be self-transcending. But this is an arduous and precarious endeavor that can leave us disillusioned or mad, as easily as enlightened and renewed" (Ibid., p. 254).

Jarvie is one of the leading philosophers of the social sciences of our age. In the early 1980s, while engaged in an ascending career, he described himself as a professional in an unresolved identity crisis. He seems to have experienced the crisis ever since he underwent two adjacent constitutive experiences. Maurice Freedman, the late chair of the Department of Social Anthropology at the London School of Economics, rejected his proposal

N. Laor (✉)
Departments of Medical Education, Psychiatry, and Philosophy,
Tel Aviv University, Tel Aviv, Israel

Child Study Center, Yale University, New Haven, CT, USA

R. Sassower
Department of Philosophy, University of Colorado, Colorado Springs, CO, USA
e-mail: rsassowe@uccs.edu

R. Sassower, N. Laor (eds.), *The Impact of Critical Rationalism*,
https://doi.org/10.1007/978-3-319-90826-7_1

for a dissertation in the PhD program in that department. And the philosophers, Sir Karl Popper and John Watkins, approved of that same proposal for a dissertation in the PhD program in the philosophy department. He wrote his dissertation under their tutelage—in Scientific Method. Since this volume is presented to him on the occasion of his 80th birthday, at the height of a long and extremely fruitful lifework, and since he found this identity crisis important enough to be aired in an intellectual-*cum*-autobiographical essay on anthropology and the irrational, it is befitting that we ask in an introductory mode what was the crisis about and what impact it had on his rich output. Did it have a pivotal driving force in the development of his philosophical world of individual and social problem-solving and of the engagement with ideas in general?

The essays in the present volume grapple with a tour de force, namely, the ideas of a methodologist who accomplished two simultaneous feats: first, turning his personal crisis into a professional tool, and second, introducing Popperian insights in the philosophy of science writ large into particular studies in the social sciences (and of necessity in aesthetics too, as he was studying art as a social phenomenon). It is very beneficial to apply openly and critically the social domain and to the world of culture Popper's proposals, including the one to consider all activity as problem-solving (and not searching for justifications). This invites improvable explanations and ongoing rethinking. Possibly, this is why Jarvie has focused since his early studies on social and intellectual ideas also in aesthetics and practices in flux (e.g., theories of cargo cults) that are in crisis and always changing. He viewed them as challenging to and therefore offered his own criticism of the (structural-functionalist and symbolist) homeostatic theories of society as either Polanyi-style or Kuhn-style authoritarian and traditionalist theories of intellectual activity and their social organization. In doing so, he was able to challenge the orthodoxy of his day and critically examine the dynamic nature and internal logics of such mysterious, mystical, or even magical cultural phenomena. The ideas of relativism and idealism loom behind all these professional stances, and in this regard Jarvie is the champion of universalism and realism as part and parcel of any critical philosophy. From this perspective he ventured to offer a theory of social change.

Jarvie discusses the intellectual environment that rests on dogmatism and that demands conversion to these dogmas as conditions for entry into it. In Popper's view this is what renders these societies more closed than open. They both find these ideological environments hindrance to intellectual curiosity, to critical examination, and to the development of the rational spirit of scholarship and of policy-making. Hence, contrary to the prevalent trend of the day, and contrary to Popper's proposals, Jarvie entertains an unusual, dual stance vis-à-vis magic, both in his anthropological studies and in his studies of rationality. He recognizes the place of magic in all human societies and applies to it his method of study: he does not equate the admission of its very existence with the admission of its claim to be the right way to solve problems or as the only way to understand human thought, behavior, and social relations. Rather, in the spirit of the expansion of the notion of rationality, he proposes (with Agassi) to consider rationality a matter of degrees, as a spectrum that ranges from the very strict to the somewhat lax, but still within the realm of rational discourse.

Has Jarvie met his own challenge to offer a theory of social change? Our answer is in the affirmative: Jarvie improves on Popper's theory of social change. The question is, are new social problems and solutions negotiated on the way to implementing social change. While Popper focuses on problem-solving within given problem situations, he allows for but does not explain how these emerge. Jarvie does: he offers recommendations for gradual implementation (piecemeal social engineering is the phrase commonly used) of new social ideas with a careful attention to their interactions with the problem situations that give rise to them. Popper underscores the inevitability of unintended (positive or negative) consequence of such processes and therefore insists on the trial-and-error approach, acknowledging human fallibility. Jarvie, in his turn, explains how, within the framework of Popper's methodological individualism, various individuals negotiate—clash, criticize, and tentatively agree upon—a plan for social action, given the different maps that they have outlined as cognitive representations of society as a whole and the social relationships that they embody. According to Jarvie, the unintended consequences are the very subject matter of the social sciences; unlike the natural sciences where we study inanimate objects or objects with no autonomous volitions, the social sciences deal with beliefs and anxieties, feedback loops, and mirroring effects. Within Jarvie's account of the workshop rationality (2005), very much the way he encountered it in Popper's seminar. (Jarvie describes the workshop mentality practiced there not as a personal experience but as a theoretical framework.) Jarvie's accounts of social process present them as capable of being thoroughly and inherently democratic. In doing so, he offers generations of scholars and educators a method of individual behavior (and social modeling) that is at once also a central constituent of the open society Popper-style.

Jarvie's contribution has closed a circle. His idea of the workshop rationality, when implemented within a given institutional setting, allows for the democratic critical examination and change of social standards, not only of the occasional change of individual behavior. Likewise, his proposed method for anthropological study recommends viewing individual encounters with others (whether other scholars in the commonwealth of learning or the targeted groups that we observe in remote regions) as potentially challenging the very commitments with which our encounters begin; when made operationally critical, this method also serves as a meta-commitment to identity—of both actors and those who observe them. Meta-commitments (as scholarly presuppositions or as intellectual prejudices), too, may be challenged and with them, the identities of all related participants. This way, perhaps, Jarvie's professional crisis has given rise to a philosophical-*cum*-anthropological extension of the methodology of anthropology: he indicates what change it requires to fall within the purview of the workshop mentality. He does this by prescribing that students of social and intellectual problems live on the edge: always ready to make a decision to endorse a hypothetical stance vis-à-vis certain descriptions and explanations, without giving up on their universalism, pluralism, and realism. (Note that Jarvie's pluralism rests on the varieties of particular perspective of individual thinkers facing particular problems as well as on the particular social/intellectual system under study. By contrast, the pluralism of the anthropologist rests on the particular singularity of the systems they study while overlooking their universal aspects.) Jarvie's self-professed personal and professional detachment has shown up as thoughtful humaneness and intellectual involvement, a method that transcends the standard interpretation of the Popperian methodological individualism and piecemeal social engineering in that it prescribes attending to the balance between the particular and the universal in a predicament as well as between the rational and the irrational parts of reasoning within the given system. Thus, whereas Popper proposes to shun utopianism for fear of historicism, Jarvie both recommends a utopian aspiration free of historicism and the study of historicist societies.

Jarvie reports that he has started out with a skeptical personal bent and that he found in Popper's philosophy a proposal that could be applied as an exploratory tool. Such a transition from the social to the cognitive could be construed as a matter of conversion, Jarvie's stress on autonomy and individual decision-making in term of his own intellectual development notwithstanding. Could one's intellectual choices within a given social milieu transcend one's personal bent? This question has imposed itself on Sigmund Freud. His answer to it was in the negative. He dismissed all philosophy as but a projected metapsychology. Whether a social philosophy is a projection or an excuse for anything, Jarvie suggests we ignore its origin and the intention behind it and try to take it literally, so as to judge its merits and defects as impersonally as we can. He ventures to underscore the reciprocal critical contribution of individuals and their culture. Could Jarvie have chosen a different theory of critical rational bent? In at least one essay (in collaboration with Agassi) the answer is that it boils down to the choice among various lifestyles. Jarvie's claim is that commitments and beliefs ought to be entertained critically and thus always tentatively; but they should not be ignored by social thinkers or deemed irrelevant to their social investigations.

Let us note that the ideal of covering all options within social theory is universal. Tradition rejects all false hypotheses. Popper suggested social researchers offer hypotheses to explain social phenomena. These may fail to explain. If they do, then they may be false yet untestable; he suggests we try to render them as highly testable as possible. Refuted hypotheses have no room in classical methodology but are significant for Popper. A hypothesis that signifies in Popper's system ascribes full rationality to individuals; however, Jarvie allows for a partial rationality. This makes his system the first inclusive system: it allows for more hypotheses than any other system, and if testable, then they count as empirical, and as such it invites social research to cover more facts than any system before it.

Jarvie's style of intellectual and social involvement has benefitted us all, from his role as managing editor of one of the most erudite and critical journals, *Philosophy of the Social Sciences*, to his social commitment to the Popperian community of scholars, initiating, convening, presiding, presenting, and editing volumes of papers given in our central conferences, and in his initiating and editing *Festschriften* for his colleagues (Watkins, Gellner, and Agassi). In addition, Jarvie is an avid scholar who has not left to others to explore the implications of his own ideas, and his areas of interest carry their mark. Perhaps the record of Jarvie's output explains the variety of perspectives and topics covered in the present volume. Jarvie has published 13 single-authored books, from *The Revolution in Anthropology* (1964), *Movies and Society* (1970), *Concepts and Society* (1972), *Rationality and Relativism: In Search of a Philosophy and History of Anthropology* (1984), *Thinking about Society: Theory and Practice* (1986), and *Philosophy of the Film* (1987) to *The Republic of Science* (2001); he has co-authored and co-edited 16 volumes, from *Hong Kong: A Society in Transition* (1970, with Agassi), *Rationality: The Critical Approach* (1987, with Agassi), and *Critical Rationalist Aesthetics* (2008, with Agassi) to the three-volume *Karl Popper: A Centenary Assessment* (2006, with Miller and Milford) and *The Sage Handbook of Philosophy of the Social Sciences* (2011, with Zamora Bonilla). In addition, Jarvie authored and co-authored some 61 essays, 52 book chapters, 28 film reviews, and 108 book reviews and essays. This level of intellectual engagement that covers everything from film and pornography to the methodology of science and the legacy of Popper testifies to a fertile mind, disciplined professional, and a collegial collaborator. No wonder that his own legacy is being acknowledged in a *Festschrift* of his own.

Above all, Jarvie is a creative and bold-thinker leader, modest in style and generous to colleagues and friends. His life course attests to the exemplary philosophy he has put before us. Even his critics recognize that their interlocutor is a worthy intellectual opponent, one who has left his mark on the academic landscape and the theoretical/critical appreciation of films/movies in society. And as you read the contributions to this volume, you will quickly realize how much respect Jarvie's colleagues, collaborators, former students, and academic fellow travelers have for him. They engage his work and ideas critically (which is, of course, the Popperian mark of respect), and they credit him with keeping a tradition alive: critical rationalism. What Jarvie (together with Agassi) managed to do is transform the debates of the Vienna positivists (of the period between the Two World Wars) all the way into the twenty-first century. And as we have all observed in the past five decades, while Popper's intellectual stock has risen, that of the others has not fared as well. The impact of this undertaking, as we have tried to explain, is not exclusively based on the merits of the methodology but has much to do with personal humility, intellectual integrity, and professional courage.

References

Jarvie, Ian C. 1981. Anthropologists and the Irrational. Reprinted in Jarvie, Ian C. 1986. *Thinking About Society: Theory and Practice*, Boston Studies in the Philosophy of Science, No. 93, Dordrecht: D. Reidel, pp. 233–256.

———. 2005. Workshop Rationality, Dogmatism, and Models of the Mind. In *The Mind as a Scientific Object*, ed. David M. Johnson and Christina E. Erneling, 471–486. Oxford: Oxford University Press.

PART I

Philosophy of the Social Sciences

CHAPTER 2

How Should Social Engineers Develop Critical Social Science?

Nimrod Bar-Am

Seemingly Unrelated Questions

Joseph Agassi and Ian C. Jarvie are acknowledged pioneers of the philosophy of technology. Half a century ago, they contributed to one of the earliest and most significant achievements of that field: securing its theoretical independence from the philosophy of science. Pure science and engineering, they convincingly argued, are distinct practices, approached by means of distinct methodologies and in light of diverging given standards of success. In particular, they explained, technological control is often achieved and improved by means of deliberate application of refuted principles and models, whereas scientific explanations ceaselessly seek unrefuted ones. Since notions of science and technology were then (and still are quite often) confused and conflated, and since Cold War governmental budgets were by that time shaping the course of academic fields, determining their fate, effort to secure such an incisive distinction was (and still is) a valuable public service carrying significant practical implications.

Let me return in brief to the arguments and conclusions that Agassi and Jarvie have offered. I do so here not because my elders require my aid in

N. Bar-Am (✉)
Rhetoric and Philosophy of Communication Unit, Communication Department, Sapir Academic College, Shaar Hanegev, Israel

R. Sassower, N. Laor (eds.), *The Impact of Critical Rationalism*,
https://doi.org/10.1007/978-3-319-90826-7_2

recapitulating their message, but because it will allow me to highlight and suggest an interesting problem that may seem at first entirely unrelated and with which Jarvie has wrestled repeatedly throughout his long and fertile career: that of the possibly distinctive character of the social sciences, in comparison (and perhaps even in contradistinction) to the natural sciences. Since Popper urged social scientists to adopt the practices of modest social engineering in more than one respect, and even encouraged them to make use of patently false principles such as those on which modern economic theory is founded, the detailed distinctions between the scientist and the engineer that Agassi and Jarvie helped clarify may have sharpened Popper's message to the point where it invites the question of their dissent.

Science vs. Technology

The commonest blunder regarding technology, and one of the first to have been clearly noted by writers on the subject, well before Agassi and Jarvie have entered it, is its surprisingly persistent identification with applied science.[1] Let me set it quickly aside and return to the question of its persistence toward the end of this section: most of our technologies have little or nothing to do with science. Skills and crafts (techne) are acquired by exploration, practice, and imitation. Applied science, on the other hand, is the premeditated deduction of practical results from a well-formulated, testable theory. The Paleolithic species *Homo habilis* (literally "skillful man") was named so because it was believed that it commanded a distinctive set of practical skills that made it markedly distinct from its ancestors. Today we know this to be an oversimplification, since some of his evolutionary predecessors seem to have acquired a number of those skills (as have modern-day chimpanzees). But the general lesson to be learned here is obvious: technology made us who we are. It therefore almost always precedes science. Applied science, in contradistinction, and by definition, does not.

As many of the early writers on technology noted, it is quite natural that science should falter and totter after our crafts and our skills. Scientists aspire for refutable explanations with ever-growing precision and generality, and these often come with a price: precision, to name one, typically comes at the expense of successful action. Many of our behavioral skills, to take a trivial example, were acquired by tacit internalization of the behavior of others, as Ryle (1949) and Polanyi (1958) famously noted, and rarely if ever have they been improved by a formal and rigorous deduction from a general and abstract theory of behavior. Were we to consult such a theory on our every move, in the hope of improving our behavior, we would quickly collapse.

The terrific metallurgist and historian of science Cyril Stanley Smith reminded us what an impressive variety of subtle manipulations of copper, iron, tin, lead, brass, bronze, and other metal alloys and solders have been known, practiced, and improved before any reasonable theory existed that could significantly contribute to these manipulations (Smith 1967, p. 59). Virtually all of the techniques of handling iron by means of fire that accumulated until the end of the eighteenth century, he noted, had been selected because they unknowingly, and yet often highly successfully, controlled the amount of carbon in it.

This is not to deny that in the past two centuries scientific knowledge increasingly contributed to the improvement of our technologies. It goes without saying that the pure scientist enriches the engineer's toolbox with computational tools as well as exploratory frameworks. But even today the general rule seems to be that exciting technologies are explored and tested mostly as improvements of existing technologies rather than as new and insightful applications of recent scientific insights as Derek Price (1968) noted. And it should be noted that the engineer often generously reimburses the pure scientist by hitting upon new challenges, by discovering unexpected puzzles, and even by suggesting original heuristic directions that could expand the pure scientists' metaphysical outlook. The atomic race, to take a worn out example, has had well-documented effects upon the development and popularity of the big bang theory.

And yet technology and science, although undeniably intimately entangled, clearly advance by means of distinct dynamics, in virtue of applications of strikingly divergent research methods, aiming for different goals (and in light of markedly dissimilar ideals). This was the point stressed by Agassi and Jarvie. The pure scientist seeks truthful explanations of reality, they observed, whereas the engineer seeks effective control over it (Agassi 1974, 1985; Jarvie 1974). This difference is of considerable significance, they added, because control can often be achieved with little or no understanding of underlying principles, and sometimes even by deliberate acceptance of false ones. Some parameters for progress adhered to by the scientist and the engineer converge: both the scientist and the engineer seek precision, clear problem situations, and minimization of surprise. But they diverge in almost all other respects: the engineer, for example, seeks improved durability, reliability, comfort of use, and reduction of production costs, while the scientist seeks increase of explanatory power, simplicity, generality, and abstractness.

Significantly, then, unlike scientists, and with good reason, engineers focus little, if at all, on either explanation or truth. Consequently, one of the most interesting differences between the engineer and the scientist is to be found in their attitude toward refuted theories (Agassi 1985; Jarvie 1986, pp. 304–308). The paradigm case here is Newton's mechanics since it underlies more practical technologies than any other scientific theory. Indisputably, its usefulness to the engineer is greater than that of the theory of relativity, which replaced it: we made it to the moon based on Newton's theory many years after it had already been widely acknowledged as, strictly speaking, an incorrect description of reality from an exact and uncompromising scientific perspective. The scientist, then, insofar as she is a strict and steadfast seeker of the truth, abandons the refuted theory when it is refuted and because it is refuted. As a scientist, she should not settle for mere approximations. As an engineer, however, she would often prefer the refuted approximation to its more exact and refuted alternative. She would do so for at least two good reasons. First, as already noted, it is often easier to plan one's moves and anticipate their results with the help of a simpler, even if refuted, theory. This is undoubtedly the case if we compare Newton's theory to Einstein's theory, but it is also true generally for the rather prosaic reason that the way to show any theory to be a mere approximation and to improve upon it typically proceeds through improving the resolution and accuracy of its description, and this, in turn, almost always increases the complexity of the description of phenomena significantly and sometimes even requires a new level of description. Second, refutation allows the engineer to know the precise limits of successful application of the technology that she uses. The limits of application are clearer in refuted theories exactly because they are refuted. These limits may be still unknown in artifacts that have been designed and built in light of theories that have not yet been refuted and which may well someday be refuted in ways and in contexts that have not yet been tested or even imagined. (And so the mere word "approximation" seems to suggest to us an explanation of the curious durability of the technology-and-applied-science blunder: it is not easy to transcend it because applied state-of-the-art science is no less applicable, and sometimes more applicable, once it becomes applied refuted science and so, strictly speaking, technology.)

The last but by no means least item that I wish to mention in this short summary of a vast literature on the differences between science and technology is ethics. Pure science, Agassi and Jarvie argued, is more or less a platonic effort and so is either morally neutral or very near so. Technology, on the other hand, by its very nature as an enhancement of our powers and our options is not; it cannot be considered as morally unbiased or inconsequential. Every choice the engineer makes thus is rife with social implications, and these are charged with moral repercussions. The mathematics and physics of projectiles or ballistics, for example, are more or less morally neutral. It can be used to study the success and failure of imaginary baseball throws. Not so its application in the service of the weapon industry. Consequently, pure scientists seem morally guileless or near so, but engineers as well as their supervisors are, and should be, legally accountable for their acts. A world which can be annihilated by a mere push of the button has a fundamentally different moral makeup from that same world before such an act was available: its inhabitants carry moral responsibilities and burdens that their predecessors did not and could not possess (Agassi 2003).

Natural Science vs. Social Science

Having set the record straight in the matter of Agassi and Jarvie on the science/technology distinction, we can turn to the matter of Popper and Jarvie on the social science/natural science distinction. This, however, is by far the more difficult and delicate task. Popper's views on this matter are all too often shrouded by the thick fog of his outstanding battle against historicism. Unlike Jarvie, Popper did not take up the theoretical question of the possible uniqueness of the social sciences directly, aside from his polemics with well-chosen archenemies of the open society. This point is made as a cautionary note to what comes next, for we are entering a treacherous hermeneutic ground. In particular, polemic contexts may easily mislead those who are hoping to extract from them neat theoretical frameworks which they were not designed to express. Indeed, they invite various rhetorical biases, most notable of which is overstatement of one's dissent by means of excessively bold demarcation contours and conceptual dichotomies. I have observed elsewhere what a high price Popper had paid when, for the sake of accentuating his dissent from his positivist interlocutors, he offered the regrettable and misleading assertion (which he himself so successfully condemned elsewhere) that certain theories (Marxism and

Freudian psychology) rather than certain attitudes (dogmatism and defensiveness) can be targeted by means of his demarcation criterion (Bar-Am 2014, pp. 693–696). Attempting to extract and reconstruct Popper's view regarding the differences between the natural and social sciences from his polemical masterpieces, *The Poverty of Historicism* (1961) and *The Open Society and Its Enemies* (1945), offers similar opportunities for confusion of the rhetorical with the philosophical, by reader and writer alike.

Consider the classic Section 27 of *The Poverty of Historicism* (1961, pp. 105–119). Popper presents there a polemically powerful distinction between trends and laws. They are, he argues, *radically different things* (pp. 115–116; italics in the original). Physical laws, he explains, are genuinely universal statements, whereas historical trends and social generalizations (even biological law-like statements such as "All vertebrates have one common pair of ancestors") are not. Despite their logical appearance, such historical generalizations are in fact existential statements expressing particular matters of fact, he concludes. This neat division between matters of fact which (by their nature? by accident?) seem to allow genuine law-like generalizations and those historical ones which (by their nature? by accident?) seem too particular (too deeply involved with the specifics of "initial conditions") to be explainable by genuine and neat law-like generalizations facilitate Popper's formidable attack against historicists, like A. Toynbee, who were attempting to derive (by some form of crude historical induction) long-term social predictions from historical progressions of events demonstrating at most historical trends. But it may also entice readers to inquire further about the general principles underlying this fine critique: Can there be social laws equivalent to Newton's law of inertia? Can there be genuine law-like (nontrivial) universal explanations within the social sciences? If yes, where are they? Is the law/trend distinction useful for distinguishing the physical from the social? (Note that Aristotle's contention that 'all earthly objects that have been thrown into the air will fall back to earth' should be regarded as an existential statement a la Popper, and yet it was clearly intended by Aristotle as a law of nature.) And if no social equivalents of Newton's laws can be formulated, why not? What is the metaphysical reality that underlies this curious situation? Should we allow metaphysical stipulations and/or empirical conjectures to influence our logical analysis of statements? Are vertebrates and societies incapable of participating in genuinely universal explanations? Are "force" and "space" more general somehow than "custom" and "vertebrate"? How are we to recognize the seemingly universal from the genuinely so, the genuinely general law from the (accidental) trend?[2] I could not find unequivocal answers to these questions in Popper. Polemically he may well be guileless: these fundamental questions are external to his immediate task which was critical. But, following Jarvie, readers may seek more than polemical guilelessness: they may seek the critical rationalists' philosophy for the social sciences.

It is for this reason that I regard Jarvie's attempts to reconstruct "Popper on the differences between the natural and the social sciences" (Jarvie 1982) as an admirable achievement. It is Jarvie's characteristically humble manner of politely inviting us to follow him beyond his mentor's shadow, without loudly boasting his originality. (Jarvie is easily the gentleman among us critical rationalists.) His mere acceptance of that task invites discussion of various possible ways of sharpening Popper's message, and so unavoidably he leads us to the delicate question of their possible dissent. When Jarvie generalizes his interpretation of Popper's polemical remarks into a philosophy of the social sciences, we are already treading in philosophical terra incognita. To take a fleeting, but by no means minor, example: unlike Popper (and following Agassi), Jarvie is unmistakably clear about his rejection of the nature vs. convention dichotomy (1986, pp. 7–10). In light of this clarification, the problem situation of seeking the uniqueness of the social realm in possible contradistinction to the physical realm has already been improved profoundly.

And it is here, I suggest, that Jarvie's outstanding inquiries into the methodological status of technology are invaluable, for they sharpen the close, many-faced, and sometimes rash analogy that Popper seems to have drawn (and recommended) between the social scientist and the engineer (often in contradistinction to the "pure" physical scientist). Popper, to be clear, had no explicit philosophy of technology. And so Jarvie's patient inspection of the differences between the engineer and the scientists invites open discussion of the theoretical limits of Popper's analogy. Why, asks Jarvie, should social scientists settle for piecemeal engineering when physicists freely discuss the state of the universe billions of years into the past or the future? (1986, pp. 328–351). Clearly, the fact that social scientists cannot justify long-term predictions by means of observed social regularities is irrelevant here, since justification is impossible anyways. Shouldn't utopian engineering, then, openly clear and critical about its known methodological limits, be just as acceptable as long-term physical predictions?

The most striking and distinctive feature of social explanation, Popper and Jarvie seem to agree, is that it presupposes goals, and goal-directedness. This feature is sharpened and clarified by Jarvie's discussion of the uniqueness of technology. For, as we have noted, in contradistinction to purely theoretical physical science, engineering oozes with moral repercussions and responsibilities. Does this extend to social science, then? Certainly. But, can we not have social theories that are ethically neutral? Can we have social scientists that are ideology free? If not, how, if at all, should this influence the methods of social science? And how should we best approach immoral practices? Popper touches upon these crucial questions in brilliant footnotes of his *Open Society*. But Jarvie puts them at the center of our attention in and in unprecedented resolution. And since he finds it difficult to imagine that Popper intended to suggest that social scientists abandon their search for truth, for the sake of some form of pragmatic social control over short-term social phenomena, he endeavors to improve the obviously false principle of rationality (Agassi and Jarvie 1973).

Conclusion

I would like to conclude with a brief word of criticism. A reader's complaint really. Popper and Jarvie share an awkward observation about the uniqueness of the social sciences whose significance (outside of criticizing inductivists) is unclear to me and to the best of my knowledge was never properly clarified by them. My own impression is that it is either a reminder of positivistic speech habits (which is at times noticeable in Popper, though quite rare in Jarvie) or an expression of metaphysical absentmindedness or naiveté. I am referring to their claim that social phenomena (and thus social research) are somehow more involved in the presupposition of "initial conditions" than their physical parallels (Jarvie 1982). For Popper and Jarvie, this means that social phenomena are supposed to be somehow more particular and thus more accidental, perhaps even more chaotic in some subtle sense, or otherwise less liable to be successfully generalized in some sense that is intimately connected to the trend/law distinction, and so its metaphysical status is nebulous at best. Here, as elsewhere, Jarvie's explication of the social situation by means of its translation into the context of technology is extremely helpful, but it is helpful not just in sharpening his main point (that methodologically, this difference is minor) but also in exposing the problematic status of this contention.

It is with much commonsense that Jarvie explains it by noting, for example, that transportation and shelter challenges or food-seeking tasks around the world are far more context dependent than general speed and distance calculations. And, he adds, since all technological artifacts are bounded by specific demands for precision, cost limits, and various time-frame restrictions, technology is more of a parochial affair than general physical quests. It is very difficult to comprehend what this contention amounts to, metaphysically speaking. Does it imply some general hierarchy of scientific abstractions? Is such a hierarchy implied by the trend/law distinction? Are abstract supply and demand equations less abstract than Newton's laws? Most importantly, is the science of attempting to explain the success and failure of engineers different somehow from the science of attempting to explain the physical nature of our world? A clarification of this point would prove helpful for the future development of critical rationalism.

Notes

1. See, for example, the interesting early collection of essays *Applied Science and Technological Progress*, published by The National Academy of Sciences in 1967, and especially C.S. Smith's paper discussed in the following pages. See https://books.google.co.il/books?id=BTcrAAAAYAAJ&printsec=frontcover&dq=inauthor:%22National+Academy+of+Sciences#v=onepage&q&f=false.
2. Agassi (1971) argues that metaphysical frameworks render some generalizations of natural laws and other uninteresting trends. Would Popper concur? Let us assume that he would. Would this have led him to reformulate his abovementioned Section 27 of *Poverty*? It is extremely difficult to answer these questions.

References

Agassi, Joseph. 1971. What Is a Natural Law? *Studium Generale* 24: 1051–1066.

———. 1974. The Confusion Between Science and Technology. In *Contributions to a Philosophy of Technology*, ed. F. Rapp, 40–60. Dordrecht, Holland: D. Reidel.

———. 1985. *Technology: Philosophical and Social Aspects*. Dordrecht, Holland: Kluwer.

———. 2003. The Impact of Auschwitz and Hiroshima on Scientific Culture (in Hebrew). In *Conversation and Discourse*, ed. Noa Naaman-Zauderer and Yaron Sanderowitz. Tel Aviv-Yafo: Tel Aviv University.

Agassi, Joseph, and Ian C. Jarvie. 1973. Magic and Rationality Again. *The British Journal of Sociology* 24: 236–245.

Bar-Am, Nimrod. 2014. The Structure of Scientific Revolutions, 50th Anniversary Edition. *Philosophy of the Social Sciences* 44 (5): 688–701.

Jarvie, Ian C. 1974. The Social Character of Technological Problems. In *Contributions to a Philosophy of Technology*, ed. Friedrich Rapp, 86–92. Dordrecht, Holland: D. Reidel.

———. 1982. Popper on the Difference Between the Natural and the Social Sciences. In *In Pursuit of Truth*, ed. Paul Levinson, 83–107. New Jersey: Humanities Press.

———. 1986. *Thinking About Society: Theory and Practice*. Dordrecht, Holland: D. Reidel.

Polanyi, Michael. 1958. *Personal Knowledge: Towards a Post-Critical Philosophy*. London: Routledge and Kegan Paul.

Popper, Karl R. 1945. *The Open Society and Its Enemies*. Vol. 2. London: Routledge.

———. 1961 [1957]. *The Poverty of Historicism*. London: Routledge.

Price, Derek. 1968. *The Differences Between Science and Technology*. Thomas Alva Edison Foundation.

Rapp, Friedrich. 1974. *Contributions to a Philosophy of Technology*. Dordrecht, Holland: D. Reidel.

Ryle, Gilbert. 1949. *The Concept of Mind*. London: Hutchinson.

Smith, Cyril S. 1967. A Historical View of One Area of Applied Science—Metallurgy, *Applied Science and Technological Progress*, The National Academy of Sciences, 57–71.

CHAPTER 3

On Economic Methodology Literature from 1963 to Today

Lawrence Boland

Introduction

Since the famous 1962 American Economic Association meeting's session on the "Problems of Methodology," economic model builders have rarely if ever talked about methodology in their published research. This is particularly so in the case of research journal articles. The question to be considered here is why this is so.

Obviously, I am not the first to try to explain why. Part of a 2001 issue of the *Journal of Economic Methodology* was devoted to explaining why. Bruno Frey (2001) argued that the disregard is due to careerism: promotion and tenure process and the refereeing process. Both demands leave little room for discussion of methodology. Wade Hands (2001) argued that methodology is of no interest to scientific researchers who think methodology provides rules. Other than the rule concerning testability, there are no rules in published economic methodology. In a 1994 issue of that journal, Tony Lawson (1994) argued that the reason for the disinterest in methodology is because economists think methodology is rooted in

L. Boland (✉)
Department of Economics, Simon Fraser University, Burnaby, BC, Canada
e-mail: boland@sfu.ca

R. Sassower, N. Laor (eds.), *The Impact of Critical Rationalism*,
https://doi.org/10.1007/978-3-319-90826-7_3

positivism which they reject, although it is not clear what he meant by "positivism." Stavros Drakopoulos (2016) argued that it is because economists consider the scientific ideal of physics to be the appropriate methodology for economics, and so there is no need to consider the views of economic methodologists. My argument will be similar to Lawson's but more specific to the behavior of analytical philosophers who dominate North American philosophy departments.[1] But before arguing this, we need to consider the history of economic methodology since 1962.

A Brief Prehistory of Economic Methodology

At the session on "Problems of Methodology," six economists and one philosopher participated.[2] The session has become famous mostly for Paul Samuelson's discussion of Milton Friedman's view of methodology. But for ten years after that one event and its publication, research in the methodology of economics was basically banned from being published in any major economics journal until briefly in 1973. That year a major methodology article was published. That article was by my former student, the late Stanley Wong, and was about Samuelson's view of Friedman's methodological views.[3] Until 1973, no major journal published any article about economic methodology. And after Wong's (1973) article, there was none until 1979. Books fared a bit better as there were four books on economic methodology published, but none gained notice until Mark Blaug's (1980) methodology book. After that, books specifically on methodology began being published by major publishers.

After Wong's 1973 article, the next major methodology article published was probably my 1979 article in the *Journal of Economic Literature* that presented a critique of most of the criticisms of Friedman's methodological pronouncements and concluded by presenting my brief critique which identified Friedman's methodology as merely a form of eighteenth-century instrumentalism. Despite the publication of Wong's article in 1973, the editor of *The American Economic Review* continued to reject articles on methodology, including my subsequently published 1981 article about the neoclassical maximization hypothesis. It was subsequently published only after a new editor, Robert Clower, took over the editorial duties.

On the Hijacking of Economic Methodology

The history of economic methodology can be a little confusing since it is not clear what is meant today by "economic methodology." In this regard, 30 years ago Deirdre McCloskey raised a distinction between what she called "small-m" methodology and "big-M" methodology which suggested a difference between what methodologists like me talk about and what philosophers of economics talk about. What I have been doing for over 50 years is trying to address a simple question: Why do economic model builders assume what they assume? I once called my version "metatheory," since it really involves theorizing about the process of building theoretical explanations and models. In contrast, philosophers of economics worry about questions such as: What do economic model builders mean by "realism" or "realistic" assumptions? Are economic models testable? What is the cognitive status of economic theory? Is economics a science? And so on.

Needless to say, there is a considerable overlap between these two approaches, but they are different. While small-m methodologists will often be aware of the views of many philosophers, few will have studied philosophy. Similarly, most big-M philosophers of economics rarely know much about economics beyond the level of Economics 101. And when they do look into the writings of economists, they are doing so with an interest compatible with philosophical studies that usually involve what Don Ross (2014) calls "conceptual analysis." In North America, this is thus limited to the interest of analytical philosophers.

Most big-M methodologists today think any discussion of methodology must be in terms of the views accepted in mainstream philosophy departments. However, among those discussing economic methodology, there is rarely any discussion about the views of philosophers of science other than Karl Popper's. Today, even those debates have disappeared. While the philosopher Daniel Hausman in the 2001 issue of *JEM* argued that there is now a merging of these two views, to the extent that he is correct, I see this rather as a hijacking of small-m methodology by analytical philosophers.

One main question to be discussed in this chapter is why today's mainstream North American economic model builders still do not say anything about methodology in their published research. Another question to consider is why those attending methodology conferences say little about the small-m methodological decisions that model builders nevertheless must silently make. Given my long history of published books and journal articles explicitly on small-m methodology of model building, my discussion will for the most part be from a first-person perspective.

The History of Economic Methodology Since 1980

Fortunately, methodology discussion has been active in history of economic thought conferences since the early 1980s. For the History of Economics Society, its inclusion was deliberate and explicit, thanks to Warren Samuels and Mark Perlman. And then, of course, like many subdisciplines of economics, the need to publish for the demands of tenure and promotion eventually led to the creation of journals devoted to the philosophy and methodology of economics. The first was *Economics and Philosophy* which was first published in 1985. Four years later, the first journal explicitly devoted to economic methodology, *Methodus*, was created in Hong Kong for an organization explicitly devoted to economic methodology: the International Network for Economic Methodology (INEM).

From its beginning, this organization had over 80 founding members (the list of founding members is published in every issue). Among the founding members there are very few philosophers. In mid-1994 *Methodus* was renamed the *Journal of Economic Methodology*, and today it continues publishing economic methodology. It appears that the original editors of the JEM lost control of the journal in the mid-1990s, and some think this renaming event was really a hijacking by a group of historians of economic thought sympathetic to the big-M view of methodology. In particular, this group opposed anything dealing with Popper that did not accept the view of Popper's philosophy that was created by Imre Lakatos. It is Lakatos who invented the idea of a "Popperian falsificationism" and the view of Popper as being a "naive falsificationist" (the phony view that claims scientists are only interested in testing and refuting theories and his philosophy models).[4] Popper explicitly rejected Lakatos' falsificationist characterization of science.[5] Nevertheless, it persisted in economic methodology discussions mostly because of Blaug's promotion of Lakatos' view. INEM continues to hold conferences, but today it, too, has been hijacked by interests of analytical philosophers. Those few of us still interested in small-m methodology now find little of interest in INEM meetings.

INEM has been holding meetings about methodology roughly every other year, beginning when it held them usually in conjunction with the History of Economics Society's meetings. Recent INEM conferences were held in Cape Town, South Africa (2015), and in San Sebastian, Spain (2017). And if there is any doubt about my claim that philosophers of economics have hijacked the INEM, one need only consult their call for the last conference where it says contributed papers "are welcome in all areas of philosophy of economics, including the methodology of economics." Apart from the INEM conferences, there have been several independent conferences held to discuss specific topics, such as Popper's or Friedman's view of methodology. (One was a 1985 conference on Popper held in Amsterdam; most of the conference's papers were published in a book which I reviewed in 1990.[6]) Despite my well-known publications in economic methodology, for some unknown reason I was deliberately not invited. This was surprising since, by most standards, I was the only Popperian methodologist publishing about economics.

The second methodology conference was held in December 2003 to celebrate the 50th anniversary of the publication of Friedman's famous essay on economic methodology, the exact same essay I wrote about in my 1979 *JEL* article discussed above. Despite my 1979 article being one of the most cited methodology articles, I was not invited to the methodology conference. Most of the papers presented at that conference were subsequently edited and published by a prominent analytical philosopher of economics in a 2009 book which I reviewed for *Economics and Philosophy* in 2010.

The reason I note these two conferences and their published essays is not only that they were explicitly about economic methodology but also because they illustrate a sickness among philosophers and some economists interested in the history of economic thought and methodology. Why would someone organize a conference on Popper's legacy in economics and not invite someone who is prominently known as a Popperian methodologist? Similarly, why would someone organize a conference about Friedman's famous methodology essay and not invite the author of the most cited methodology article about that essay? I raise this not because I happen to be the victim of this sickness but as it illustrates something I have often seen at methodology conference sessions where someone in the audience stands up and criticizes someone who is not present (or deliberately not invited).[7] I find this behavior unethical. It also displays a weakness in the confidence many philosophers of economics and some historians of economics interested in methodology have in their own views of economic methodology.

On Why Economic Model Builders Have Rarely Been Interested in Methodology

Some may think I am wrong about economists' disinterest in methodology since all during the 1940s, 1950s, and 1960s model builders were often talking about the testability of their models. But, the testability rule often discussed (particularly when I was a PhD student in the 1960s) was never based on methodology. It was the foundation of a response to critics of the growing use of mathematics in model building in the late 1930s. The critics claimed that using mathematics will only produce tautologies. And given this criticism, Samuelson set out to prove the critics wrong in his PhD thesis which he started writing in the late 1930s. He subsequently published it as *Foundations of Economic Analysis* (1965) after the end of World War II. Interestingly, an early version of that book was titled "The Operational Significance of Economic Theory." Starting in the beginning of the book, he makes clear his interest in what most would consider the testability of mathematical models. But the primary issue is refutability. The problem with a tautology is that it is not conceivably false, hence irrefutable. My usual example of a tautology is the statement "I am here or I am not here" which is true regardless of where "here" is or who "I" am. Samuelson's way of addressing this was to talk about what he called "meaningful theorems" which simply are non-tautological statements. As he says: "By a *meaningful statement* I mean simply a hypothesis about empirical data which could conceivably be refuted, if only under ideal conditions. A meaningful theorem may be false" (1965, p. 4). In all of his discussion of economic methodology, nowhere does he refer to a philosopher. This is also true of Friedman's famous 1953 methodology essay.[8] Economists rarely if ever discuss the views of philosophers when discussing what some usually consider matters of methodology.

At the beginning I discussed various reasons offered to explain why economic methodology is not of interest to economists. There was careerism, methodology as rules, rejection of positivism, and the view that the appropriate methodology is already evident in the scientific ideal of physics. My explanation instead involves the behavior of philosophers, one that many economists may have encountered when discussing their research with philosophers. The behavior I have in mind is what I encountered in a wonderful special weekly faculty seminar which we called our Pi-Digamma Seminar. In this seminar the attending faculty would take turns presenting and discussing their research. We had faculty from various science, social science, and even from humanities departments; in one seminar our university president presented his research on the history of China. One of the attendees was even a member of the philosophy department. His attendance became a problem because he would interrupt the presenter telling them that they were wrong about this or that. Too often it was a disagreement about a word or concept the presenter was using. Eventually, the seminar attendees asked the philosopher to not attend. After that, the seminar got much better.

What the philosopher displayed was the common attitude that philosophers too often display: a monopoly on clear thinking. They impart this not only to fellow faculty but to their students, too. This common attitude is the cause for economic model builders to avoid engaging in questions that philosophers might likely challenge. Even though small-m methodology questions rarely involve topics about which philosophers have any expertise, they think they do. As I said above, the testability that concerned early model builders had nothing to do with what philosophers, such as Popper, talked about. It was just a simple matter of the logic that the critics of early mathematical model building did not understand. It was not something that required an exercise in analytical philosophy.

My Ignorant Start in Methodology

I began doing research in methodology as a result of my first rejected PhD thesis on capital theory. My beginning was due to my following the instructions of the thesis committee. Fortunately, one of the committee members was a teacher of one of my classes where I criticized what the teacher was putting on the blackboard. My criticisms were usually correct, so he was not eager to dismiss my effort to obtain a PhD. He also was a graduate student at one of the Ivy League schools where he saw PhD committees easily failing students under similar circumstances and said he was not going to let it happen to me. I guess my rejected thesis must have mentioned Popper so the committee came up with the idea of writing a thesis that would employ Popper's *The Logic of Scientific Discovery* (1934) and apply it to economic model building.[9] The PhD program I was enrolled in was not your usual 1960s program. It was one financed by the US National Defense Education Act, and was designed to promote high-tech mathematical model building.[10] Almost all of our courses were high-tech mathematics oriented. We were also required to take a couple classes in the history of economic thought, but, as was usually the case until the 1990s, the history considered stopped at the middle of the 1930s with the "Keynesian revolution." So, I learned nothing about how economics had changed after WWII with the work of writers and model builders, such a Samuelson and Kenneth Arrow. Our studies often were about the work of the famous general equilibrium model builder Léon Walras and his famous 1874 book, *Éléments d'économie politique pure, ou théorie de la richesse sociale.* Needless to say, we never read that book. What was worse is that we just came away thinking Walras' mathematical model building dominated economics from then up to at least the 1960s. The history of economic thought classes did not correct our misperception. We should have learned there that institutional economics dominated economics from the end of WWI until the mid-1950s. Institutional economists rejected theoretical model building and instead were in favor of analysis of institutional data about such things as the national income accounting.

Thanks to my PhD education, I was ignorant of what ordinary economics might think about research in economic methodology. As a result, until my first publication in 1968, I had 55 rejections![11] And worse, as many of my students complained, my first two papers were published in an apartheid journal. What the students did not understand is that the journal, *South African Journal of Economics*, was published by probably the last anti-apartheid university. The first publication was a chapter from my PhD thesis. The second was the result of advice I got from Robert Mundell, the editor of the *Journal of Political Economy*. I must have submitted four papers about the methodology of model building, all of which he rejected. But he kindly recognized my ignorance and offered advice. Specifically, he suggested combining all four papers into one paper about model building. I followed his advice.[12] My next paper was published in 1970 by an editor of last resort, Geoffrey Harcourt.

Eventually, Ian became another editor of last resort. I think I first met Ian at the 1967 meetings of the Society for the History of Technology (SHOT) that were held in Toronto. I overheard him say there to my former teacher, Joseph Agassi, that he did not have a high opinion of my abilities to do economic methodology. After my next two papers were published in *Philosophy of Science*, he seemed to have changed his mind and published one of mine, the one I wrote for the SHOT meetings in Toronto. Subsequently, Ian has helped me in many ways, including giving me the opportunity to review books on economic methodology as well as submitting papers on that topic. This help has been much appreciated.

Notes

1. It may not be limited to North America if Don Ross (2014, Chap. 1) is right.
2. The session was published in the 1963 *Papers and Proceedings*, where there are papers by Fritz Machlup, Ernest Nagel, Andreas Papandreou, and Sherman Krupp, followed by discussions by Chris Archibald, Herbert Simon, and Paul Samuelson.
3. Apart from being a good methodology article, I suspect it was also published because at the time Wong was a student of Joan Robinson at Cambridge and she had been constantly criticizing George Borts (the editor of *The American Economic Review*) for never considering publishing any nonmainstream articles like Wong's.
4. Imre did not really understand Popper and created this to promote his own role in the philosophy of science.
5. See Popper's 1982 introduction to his 1983 publication of his previously unpublished *Postscript* where he says "Am I really the man who had naive falsificationism as the linchpin of his thoughts? Is the Kuhnian paradigm true? May I 'legitimately be treated as' a 'naive falsificationist,' even though Kuhn admits, after looking at *The Logic of Scientific Discovery*, that, as early as 1934, I was *not* one? .... Tests are attempted refutations. All knowledge remains fallible, conjectural. There is no justification, including, of course, no final justification of a refutation. Nevertheless, we learn by refutations, i.e., by the elimination of errors, by feedback. In this account there is no room at all for 'naive falsification'" [pp. xxxiv–xxxv] (emphasis in original).
6. The conference was held to honor the retirement of Joop Klant who was considered in Europe a prominent proponent of Popper's philosophy of science in economics. The papers were published in de Marchi (1988). My review of the conference volume appeared in the *Research in the History of Economic Thought and Methodology*, vol. 7, 1990–1992.
7. In the first case, I was chairing a HES session at its conference held in Vancouver and scolded that behavior. In the second case, I was again the victim at a session of a HES conference held at George Mason University to discuss Friedman's essay and criticisms such as mine; even though I was attending those meetings, I was not invited to respond. And similarly, at the 2003 conference I mentioned above, I was criticized by Blaug even though I was not invited to attend or respond.
8. Friedman actually met Popper and presumably discussed methodology with him; if he did, he apparently did not understand Popper's view of testing and falsifying. Friedman came away thinking that all that matters was

the refutation of predictions of a theory, and hence his rejection of the need to be concerned with using false assumptions in one's explanation.

9. Eventually, I published my thesis as Chaps. 2 and 3 in my 1989 book about the methodology of economic model building.
10. This Act financed very generous fellowships. They were so generous that I took a pay cut for my first teaching job.
11. While I think a major reason for my 55 rejects is that I was a beginning writer and not a very good one, the primary reason was, of course, the implicit ban on any major economics journal publishing articles about economic methodology that I have discussed above. As I said, I was an ignorant beginner.
12. This paper was later republished as Chap. 1 of my 1989 book.

References

Backhouse, R. 2010. Methodology in Action. *Journal of Economic Methodology* 17 (1): 3–15.

Blaug, M. 1980. *The Methodology of Economics*. 2nd ed. Cambridge: Cambridge University Press.

Boland, L. 1979. A Critique of Friedman's Critics. *Journal of Economic Literature* 17: 503–522.

———. 1989. *Methodology of Economic Model Building: Methodology After Samuelson*. London: Routledge.

———. 1990–1992. Understanding the Popperian Legacy in Economics. *Research in the History of Economic Thought and Methodology* 7: 273–284.

———. 2001. Towards a Useful Methodology Discipline. *Journal of Economic Methodology* 8: 3–10.

———. 2010. Mäki on 'Friedman 1953'. *Economics and Philosophy* 26: 376–382.

———. 2016. Philosophy of Economics Versus Methodology of Economics. *Sudia Metodologiczne* 36: 17–26.

Davis, John B. 2007. The Turn in Economics and the Turn in Economic Methodology. *Journal of Economic Methodology* 14 (3): 275–290.

Drakopoulos, Stavros. 2016. Economic Crisis, Economic Methodology and the Scientific Ideal of Physics. *The Journal of Philosophical Economics: Reflections on Economic and Social Issues* X: 28–57.

Düppe, Till. 2011. How Economic Methodology Became a Separate Science. *Journal of Economic Methodology* 18 (2): 163–176.

Dyke, Charles E. 2012. *Philosophy of Economics*. Englewood Cliffs, NJ: Prentice-Hall.

Frey, B. 2001. Why Economists Disregard Economic Methodology. *Journal of Economic Methodology* 8 (1): 41–47.

Friedman, M. 1953. *Essays in Positive Economics*. Chicago: University of Chicago Press.

Hands, D. Wade. 2001. Economic Methodology Is Dead – Long Live Economic Methodology: Thirteen Theses on the New Economic Methodology. *Journal of Economic Methodology* 8 (1): 49–63.

Hausman, D.M. 2001. A New Era for Economic Methodology. *Journal of Economic Methodology* 8 (1): 65–68.

Lawson, T. 1994. Why Are So Many Economists Opposed to Methodology? *Journal of Economic Methodology* 1 (1): 105–134.

Mäki, Uskali. 2012. *Philosophy of Economics*. Oxford: North Holland.

Ross, Don. 2014. *Philosophy of Economics*. Basingstoke: Palgrave Macmillan.

Samuelson, Paul. 1965 [1947]. *Foundations of Economic Analysis*. New York: Atheneum.

Wong, S. 1973. The 'F-Twist' and the Methodology of Paul Samuelson. *American Economic Review* 63: 312–325.

CHAPTER 4

The Republic of Science and Its Constitution: Some Reflections on Scientific Methods as Institutions

Jesús Zamora Bonilla

Introduction

One of the main virtues of Ian C. Jarvie's book *The Republic of Science: The Emergence of Popper's Social View of Science, 1935–1945* (2001) is that it convincingly demonstrates that Popper's greatest insight in the philosophy of science was to understand scientific rules as social institutions. Exploring the road from Popper's *Logik der Forschung* (1934) to *The Open Society and Its Enemies* (1945), and contrasting Popper's view of science with his view of social institutions in general, and with the views about these topics of other contemporary authors (mainly Hayek and Polanyi), Jarvie clearly shows the radical novelty of Sir Karl's approach to the problem of scientific method, as well as why the intrinsic social nature of methodological rules was difficult to discern by most readers of Popper. Another important result of Jarvie's hermeneutics is that, instead of the traditional interpretation of the young Popper's intellectual evolution, as if *The Open*

J. Zamora Bonilla (✉)
Faculty of Philosophy, UNED, National Distance Education University, Madrid, Spain
e-mail: jpzb@fsof.uned.es

R. Sassower, N. Laor (eds.), *The Impact of Critical Rationalism*,
https://doi.org/10.1007/978-3-319-90826-7_4

Society and *The Poverty of Historicism* (1944) resulted from an application of his falsificationist *methodology* to the case of the social sciences and political philosophy in general, Jarvie offers a different and more illuminating perspective: the works of the 1940s would rather be a natural development and generalization of the view of science as a *social* institution, that is, an exercise in whether society could better be understood on the *republican* pattern of science advanced in the *Logik*. I will not say too much in this chapter about this second prevailing element of Jarvie's book but concentrate instead on the first issue: the essentially conventional and hence social nature of scientific method and the reading of Popper's falsificationism as an attempted sketch of a constitution of the Republic of Science, as Jarvie magisterially tries to reconstruct, mainly in the first and second chapters of his book.

Sharing Jarvie's interpretation of the Popperian idea of scientific method, I want to discuss more specifically two concrete aspects of (in Jarvie's terms) Popper's proto-constitution of science, one which is closer to traditional debates about methodology and the other one referring to what view of social institutions would be more helpful for a project like the one Jarvie attributes to Sir Karl. The first question is about what should be the *supreme* rules of the scientific method. I will express some doubts about the idea that being permanently open to criticism has to be taken as such supreme rule, not because it is not extremely important and useful, but because I think that it is more instrumental than substantive, and has to be justified by its presumed efficacy in leading us to other goals whose expression should count as *higher-level* norms in the constitution of science (which will be the focus of the next section).

The other question has to do with Popper's view of society and of social sciences, which I will claim excessively determined and limited his vision about what kind of understanding we might reach about the idea of the scientific method as a social institution. Three quarters of a century have passed since Popper elaborated on those ideas, and both the political situation and the social sciences themselves have evolved in a direction that would justify the following conclusion: had Popper devised his social understanding of methodological norms around the end of the century, he could have chosen a very different and more appropriate set of conceptual tools to work it out. I will deal about this possibility below.

Methodological Rules as Conventions

In spite of the title of his most important book in philosophy of science, Popper's fundamental claim about the issue is that the rules of scientific research are not a matter of *logic*, in the sense that they cannot be derived from logical axioms, nor are they something like applied logic. According to Popper, the logic of scientific discovery is a logic, at most, in the sense in which we can talk about "the logic of chess":

> Methodological rules are here regarded as *conventions*. They might be described as the rules of the game of empirical science. They differ from the rules of pure logic rather as do the rules of chess, which few would regard as part of pure logic: seeing that the rules of *pure* logic govern transformations of linguistic formulae, the result of an inquiry into the rules of chess could perhaps be entitled 'The Logic of Chess', but hardly 'Logic' pure and simple. (Similarly, the result of an inquiry into the rules of the game of science—that is, of scientific discovery—may be entitled 'The Logic of Scientific Discovery'.) (Popper 2002, p. 32)

Conventions can be justified, argued in favour or against, but they cannot be proved as logical or mathematical theorems, nor can they be demonstrated or confirmed as some (other than Popperians) think the laws of empirical science can. There is a conventional element in the choice of the rules of any practice: one might decide not to play the game according to those rules, though in that case probably she will not be playing that game but a different one. Popper, hence, responds to the different forms of conventionalism (i.e., philosophical understandings of science according to which the *truth* of scientific claims is "conventional"), not by proving that conventionalism is self-consistent but by admitting that science contains an indispensable conventional component built into it; this component, however, is not the truth (or the accepting-as-true) of the theories and laws but the *choice* of the rules according to which scientists carry out the process. Nevertheless, if we want this choice not to be purely arbitrary, it is necessary that we engage in a rational discussion about the reasons why it could be *better* to play that game according to those rules instead of others, or instead of a very different game. And these reasons are necessarily related to whether the proposed rules will actually lead us to satisfy the *goals* of the game. So, before studying what the rules of science should be, we need to have an idea of *why we want* to do science rather than some different kind of activity. Just a few paragraphs after the text just quoted,

Popper clearly formulates the goal he has in mind: "the rules are constructed with the aim of ensuring the applicability of our criterion of demarcation" (Ibid., p. 33).

Popper's criterion of demarcation is falsifiability: "it must be possible for an empirical scientific system to be refuted by experience" (Ibid., p. 19; italics in the original). Or, a little bit more explicitly:

> In other words: I shall not require of a scientific system that it shall be capable of being singled out, once and for all, in a positive sense; but I shall require that its logical form shall be such that it can be singled out, by means of empirical tests, in a negative sense: *it must be possible for an empirical scientific system to be refuted by experience.* (Ibid., p. 18; italics in the original)

Hence, the rules of science, according to Popper, must be designed in order that all scientific systems (theories? models? hypotheses? theoretical frameworks?) are not only falsifiable through empirical testing but also *as much falsifiable as possible* (as it is clear from rule R5 below, and from Popper's insistence on degrees of testability throughout the book, especially Chap. 6). Before starting to discuss to what extent this is a sufficient goal on which to set a system of methodological rules, let's see what norms Popper had in mind. One discouraging aspect of *The Logic of Scientific Discovery* from a contemporary point of view is that, once its *main* goal of offering a defence of the norms of science has been stated, the book does not proceed to systematically *formulate* such rules but enters into some discussions characteristic of the philosophical concerns of the time. The lack of systematic analysis led Jarvie to gather the rules that Popper randomly formulated throughout the book. I will present here just the most important ones, starting by what Popper and Jarvie call the "supreme rule":

> (RS) The other rules of scientific procedure must be designed in such a way that they do not protect any statement in science against falsification.
>
> (R1) The game of science is, in principle, without end. He who decides one day that scientific statements do not call for any further test, and that they can be regarded as finally verified, retires from the game.
>
> (R2) Once a hypothesis has been proposed and tested, and has proved its mettle, it may not be allowed to drop out without "good reason."
>
> (R3) We are not to abandon the search for universal laws and for a coherent theoretical system, nor ever give up our attempts to explain causally any kind of event we can describe.

(R4) (Don't) use undefined concepts as if they were implicitly defined.

(R5) Only those (auxiliary hypotheses) are acceptable whose introduction does not diminish the degree of falsifiability or testability of the system in question, but, on the contrary, increases it.

(R7) Inter-subjectively testable experiments are either to be accepted, or to be rejected in the light of counter-experiments.

(R9) After having produced some criticism of a rival theory, we should always make a serious attempt to apply this criticism to our own theory.

(R10) We should not accept stray basic statements –i.e., logically disconnected ones- but ... we should accept basic statements in the course of testing theories; or raising searching questions about these theories, to be answered by the acceptance of basic statements.

(R11) Those theories should be given preference which can be most severely tested ... (i.e., favour) theories with the highest possible empirical content.

(Jarvie 2001, pp. 31 ff. Of a total of fifteen, the above are taken *verbatim* from *The Logic of Scientific Discovery.*)

The complete list of these rules is what Jarvie calls Popper's proto-constitution of science. I shall devote the rest of this section to discuss several important problems that can be recognized in this constitution, its evident lack of systemization not being one of them. A problematic question in Popper's list is whether he proposes these norms as just a *suggestion* and "invitation to play science" according to these rules or if he thinks that it is the "essence" of science to be played in exactly this way, so that those that are not following these norms are just not doing science, properly speaking. Another way of putting this question would be to ask if, for Popper, someone not following these norms would be a *bad* scientist or *no* scientist at all. Having here a set of norms (rather than a single one), and taking into account that many of them could in principle be disobeyed to a higher or lower extent (something to which I will come back later), it is conceivable that the norms were followed *more or less*, so that one could be a better or worse practitioner of Popperian science. *The Logic of Scientific Discovery*, with its insistence on the problem of demarcation, is written in such a way that a categorical reading of the methodological norms seems closer to Popper's intentions: if you don't strictly follow the falsificationist rules, you are simply not doing science, but metaphysics. But this seems too strict: one can prefer sometimes a theory with a little lower degree of testability, say, and still be a scientist, even a falsificationist scientist, as one can be a professional athlete and skip some few training days occasionally.

Perhaps it is better and more coherent with Popper's thought to interpret this "methodological constitution" as an (Weberian) *ideal type* or as a kind of (Kantian) *regulative idea*. This, however, does not totally solve the problem of whether these norms constitute a *definition* of science or have just to be taken as *suggestions* to carry out research in a certain way. As we have seen above, that methodological norms are *conventions* (and not truths) invites us to take more seriously the latter option. But this option would obviously lead to a more fundamental question: why *should* I follow those norms instead of other possible ones (perhaps slightly different, or perhaps more disparate)? Popper's immediate answer is, of course, that you should follow these norms in order to attain the goal of keeping your theories (or systems) as testable as possible. And this leads us to ask the following questions: first, would following Popper's rules, such as they are formulated, *actually* lead us to maximize the degree of testability of our theories? Wouldn't it be possible to attain this goal by some different means? And second, why should be testability the most important goal and perhaps the only goal of scientific practice? After all, testability sounds like something that is most naturally a means to attain other valuable things, something that has an instrumental value, not something that is valuable in itself. Popper often speaks in *The Logic of Scientific Discovery* of "the progress of science," "the progress of knowledge," or just "progress" as the prize attained by his recipes. Hence, why is the maximization of testability (and just testability) the only means to further the progress of science? What if the goals of science (and of scientists) are more diverse and complex and are accompanied by more labyrinthine ways of approaching them than through an obtuse fixation with testability?

Regarding the first question, a simple way in which Popperian rules could be inefficient enough is if the maximization of testability within a particular segment of the complex network of statements involved in any realistic research process could sometimes entail a significant reduction of testability in *other* segments. Perhaps in order to keep a theory highly testable we should sometimes accept other theoretical hypotheses, either more general or profound (having to do with general presuppositions of the field) or more technical ones (like those about the use of measurement devices or statistical assumptions). I doubt that the Popperian system could offer a *global* measure of testability (if developing a workable *local* measure were not utopian enough) with a clear recommendation. Perhaps Popper would insist that his norms are recommendations for the individual scientist, which has the obligation of maximizing the testability of her portion of research. But even in this case, it is an open question whether a high number of individual researchers proceeding in this way would lead science to globally maximize its degree of testability (and progress), or if there could exist some complex, negative feedback mechanisms that would better be served in a Mandevillian way (when some scientists behaved less honestly). Imagine that individual scientists could choose between developing their preferred theories following in the strictest possible way the falsificationist rules and following a more verificationist way, protecting those theories from falsification. Is there a guarantee that the first strategy would necessarily lead to a higher degree of global progress than the second one (which, by the way, is reminiscent of the Lakatosian methodology of research programmes)? Perhaps in the case of science, private vices also transform at times into public virtues.

Regarding the second question, it is well known why Popper said that he refused to mention truth as the goal of science, as he attempted to do some decades later thanks to Alfred Tarski's rehabilitation of the notion of truth without abandoning the idea that a more abstract notion of progress is what distinguishes science from other activities. One may wonder how Popper's book would have look liked if he had opted for explicitly asserting that progress towards (knowledge of) the truth is the *ultimate* goal of science. Probably, his views on methodology would not have changed substantially, for the concept of verisimilitude, or approximation to the truth, was later defined by him in such a way that a falsificationist methodology would naturally derive from or be coherent with it, in the following sense: theory X is at least as close to the truth as theory Y (according to Popper's qualitative definition) if and only if every true statement following from Y also follows from X, and every false statement following from X also follows from Y; hence, under the *hypothesis* that X is at least as verisimilar as Y, it follows that all *verified* predictions of Y will be correct predictions of X and all *verified* mistakes of X will also be mistakes of Y. (I am using the term "verified" just in the sense that these predictions have been tested and accepted on empirical grounds.) Stated differently, the hypothesis that X is at least as close to the truth as Y allows to make the second-order *prediction* that X's *tested* predictions would always be at least as good as those of Y. Hence, the provisional corroboration of this second-order prediction (that all empirical successes of Y have been *to this date* matched by X, and all empirical failures of X *known to this date* are also failures of Y) would count as a corroboration of the second-order *conjecture* that asserts that X is closer to the truth than Y. In Popper's own words:

> I do not suggest that the explicit introduction of the idea of verisimilitude will lead to any changes in the theory of method. On the contrary, I think that my theory of testability or corroboration by empirical tests is the proper methodological counterpart to this new metalogical idea. The only improvement is one of clarification. Thus I have often said that we prefer the theory t_2 which has passed certain severe tests to the theory t_1 which has failed these tests, because a false theory is certainly worse than one which, for all we know, may be true.
>
> To this we can now add that even after t_2 has been refuted in its turn, we can still say that it is better than t_1, for although both have been shown to be false, the fact that t_2 has withstood tests which t_1 did not pass may be a good indication that the falsity-content of t_1 exceeds that of t_2 while its truth-content does not. Thus we may still give preference to t_2, even after its falsification, because we have reason to think that it agrees better with the facts than did t_1. (Popper 1963, p. 235)

As it is well known, Popper's dream of supporting his falsificationist method with a formal theory of approximation to the truth received a fatal blow when it was proved that no false theory could be logically related to any other theory in exactly the way Popper's definition of verisimilitude demanded for the former being more verisimilar than the latter, and hence, the idea that we can give preference to a falsified theory over another because of *that* reason became logically untenable (more on the debates regarding verisimilitude, see Niiniluto 1998). Perhaps we could explicate the notion of a theory being closer to the truth than another in different terms than Popper's definition of verisimilitude. According to alternative definitions of truthlikeness, would the methodological norms selecting more truth-like theories be the norms Popper defended in his proto-constitution?

Instead of starting by proposing a philosophical argument about what the goal of science *should* be and logically deriving from that goal the methodological norms that could better serve to its attainment, we might perhaps follow a more empiricist or abductive procedure, examining in the first place what methodological norms scientists *actually* follow and *conjecturing* afterwards which goal or goals could they be trying to pursue such that the norms actually followed in scientific practice are efficient procedures for the attainment of *those* goals. I leave the details about this possibility for the next section; I will only mention now with respect to this question that we might find that scientists actually follow under some circumstances verificationist rules, and that this can be explained not because they are irrational or dogmatic but just because this is the best science game they can play, taking into account both their goals and the limited resources at their disposal.

The latter discussion can also be connected with a different problem. Science is (mostly) not done by algorithms, but by humans, and every person has usually *more than one goal* in everything she does. Even if the maximization of testability and the pursuit of truth (of deep, useful and interesting truths) were the only substantial goals of science as an institution, the people that carry out the activities in which science consists may and usually will have other goals and values as well. These range from diverse preferences about the epistemic qualities of the theories, models, empirical procedures, and styles of thought to more practical things, like good work conditions, access to facilities, and, most importantly in the case of many scientists, prestige and intellectual influence. Perhaps the institutional arrangement that would maximize the testability of theories is to publish everything in a strictly anonymous way, for this would probably avoid serious biases due to the publish or perish tendency of current science. But, under such institutional arrangement, it would be likely that not enough people would pursue a science career. Many scientists' thirst of glory would be hardly satisfied if science worked in such a way that no result *could* ever be declared definitively established beyond all reasonable doubt: for a seeker of scientific glory, provisional truth and fallibility are very often something we have to pay lip service to, and firm verification of a theory or law is an institutional goal more valuable than having passed till now all tests but who knows what will happen tomorrow. If scientists' motivations are close to this description, then establishing methodological rules that at least have a little verificationist flavour could be seen as a necessary price for having a powerful science system, an unavoidable element in any realistic constitution of the republic of science.

The Constitutional Political Economy of Scientific Institutions

In the preceding section I have basically considered the *content* of Popper's methodological proto-constitution of science, shedding some light on the question whether testability (and the maximization of degrees of testability) can be taken as the *sole* regulatory ideal for the normative architecture of the scientific method. In a nutshell, if testability is not seen as the only constitutive goal of science (nor as the only goal of scientists), then it is conceivable that, under some circumstances, a trade-off between the maximization of several different goals can arise, and so it may be rational to be satisfied with a little less testability in exchange for a higher level of satisfaction of the other aims. I plan to deal now with a different question, which does not refer directly to the content of the norms of science, but to what are the best theoretical tools to illuminate the process of choice or the emergence of those norms. Popper, in general, was deeply suspicious of the state of a big part of the social sciences of his time, both because of the lack of testability of many of the most popular theories and because of the ease with which many of those theories were put into ideological use. It is a pity that Popper hardly ever bothered to take into consideration the work of people like Robert Merton and his followers which might have offered an enlightening way of discussing the normative and institutional structure of science. Instead, Sir Karl seemed to have more clearly in view other approaches to the topic, like those of Karl Mannheim's sociology of knowledge (with its Marxist inspiration). We shall never know if Popper's ideas about a constitution of science would have been more explicitly developed or would have changed to some significant degree had he devoted more time to contrast his view on the topic with that of Merton's "ethos of science" (Merton 1942).

As Jarvie explains in the third chapter of *The Republic of Science* (2001), Popper's view of society, institutions, and social science is deeply influenced by the work of liberal thinkers and economists, especially his friend, and later colleague at the London School of Economics, Friedrich Hayek. As it has been argued, the influence was surely bidirectional (Caldwell 2006), but most probably it was Hayek's ideas on the structure and development of society that basically influenced Popper's view about historicism and the open society around the end of the 1930s and 1940s, while Popper's general methodology strongly influenced Hayek's epistemological views in the 1950s. In spite of Popper's subsequent deep influence on several generations of economists who tried to elaborate their theories according to falsificationist guidance (Richard Lipsey 1963), often mixed with Milton Friedman's views on positive economics, Popper's ideas about the fundamental nature and method of the social sciences didn't seem to have gone very further than what he sketched about "situational analysis" in *The Poverty of Historicism* (1944), and, in this sense, he shared Hayek's Austrian reservations about the possibility of developing a mathematical science of society. This might have been more justifiable during the 1940s when even the most well-known formalized branches of economics of the time (Hicks' or Tinbergen's macroeconomic models) were far less complex and diversified than what they turned into a decade later thanks to the work of people like Samuelson, Arrow, and Nash. In particular, the rise of game theory could have been taken as an opportunity to put more flesh on Popper's scheme of situational analysis, for it is clear (at least for some authors) that situational logic could be seen as a minimal description of the very idea of analysing social interactions by means of game-theoretic tools (see Hands 1992 for a positive view about this question and Morgan 2012 for a slightly more sceptical one). One possible reason of Popper's reluctance to advance a more formal development of his situational analysis may have been his deep scepticism about the existence of laws within the social sciences. But the more or less successful application of game-theoretic thinking to nearly all fields of social science in the last decades has not been driven by the hope of finding deep universal laws but rather by the interest in constructing a battery of alternative, useful, piecemeal models of specific types of situations, something that parallels Popper's (and Hayek's) praise of "piecemeal social engineering" over conceited attempts to manage a whole national economy, for example. Actually, a whole branch of applied game theory, known as mechanism design, could be considered as the most accomplished, state-of-the-art version of a piecemeal social engineering.

In particular, I have defended in other papers (Zamora Bonilla 2002 and 2008) that one specific branch of economics, constitutional political economy, could be particularly useful to apply the Popperian idea of a constitution of science. Constitutional political economy was inaugurated by Buchanan and Tullock's book on *The Calculus of Consent* (1962), and its fundamental idea is that we can apply rational choice microeconomic, game-theoretic thinking, to the cases in which a group of people are not just making some choices under a situation defined by the applicability of some specific norms but to those cases in which the members of the group choose the norms themselves. Buchanan and his collaborators and followers have applied this type of work mainly to study the choice of voting procedures, or of other political representation systems, and how their predicted impact on economic performance and distribution affect these types of choices. But this methodology can be applied to the choice of *any* kind of norms whatsoever. Why not, then, to the norms of scientific research?

I think an important (though unconscious) obstacle for Popper and his disciples to consider seriously this possibility was a certain distrust of the scientific community itself, particularly after the big debate of the 1960s around Thomas Kuhn's work on scientific revolutions and "normal science." If normal scientists were to choose the actual rules of scientific practice, they would probably prefer norms that made life easier for them, so to say, instead of norms enforcing the boldest and most austere search of highly testable theories and their possible falsifiers. Perhaps this possibility is neither too unrealistic nor too undesirable when we take into account that science, after all, must be practised by people organized in communities and institutions and not only by heroic geniuses. It is important to take into account that the choice *of norms* is very different from choices made *under* norms. One essential aspect of the former is that, if not only because norms are expected to be in force during a very long period and are going to affect individuals under circumstances whose specificities are difficult to foresee, norms tend to be chosen under a "veil of ignorance" (to use the Rawlsian metaphor): the choice of norms tends to be more impartial, and less subject to strategic manoeuvres, than decisions more directly related to specific circumstances, and the more general the norm, the less space for manoeuvring, in general (except in circumstances where an individual or a subset of the group occupies or expect to occupy a powerful role in the field during a long time). Hence, a constitutional choice of the most general norms of science, made by a community composed of normal scientists, would perhaps be not very different from the choice that a group of heroic or revolutionary scientists would make.

Another important aspect of the choice of scientific norms that constitutional political economy would force the analyst to justify is what the *communities* in charge of making the choice of the norms actually are: What is the relevant constituency? Is it the scientific community writ large? Would different scientific specialities or subcommunities have different methodological constitutions, depending on the availability of data, techniques, and so on? Or should the most general constitution of science be decided by the informed choice of any citizen, for, after all, the outputs of scientific research are going to directly or indirectly affect their lives, especially because of the *funding* of science? Why should the philosopher have a prominent role in the choice of these norms? If they should, would it be reasonable that the constitution of science was written and voted by an assembly of philosophers representing all the relevant methodological schools? Some of these questions are obviously wiser than others, but the very possibility of launching them opens interesting perspectives from which to consider problems not only within methodology of science but also about the social, economic, and political relevance of methodological questions, something that, unfortunately, has been absent in most of the core literature on philosophy of science during the last century.

Many of the more severe problems that science has been suffering during the last decades could be profitably approached from this constitutional point of view, for example, the replication crisis denounced in many sciences, the hardening of academic careers, the choice of open or proprietary systems of publication, and the contested role of universities in an increasingly privately funding research systems. These are very likely problems that have to do with a dysfunction of norms that perhaps had been working in an acceptable way till a few decades ago but that now should be modified if we agreed on who are the agents directly responsible for choosing the right norms. If we do not do it this way, then the invisible tentacles of a Hayekian mangle of self-organizing, emergent institutions will certainly make the choice of norms instead of us, and these norms will likely not necessarily be in *our* best interests.

Acknowledgements The author thanks Spain's government research projects PRX14-00007 and FFI2014-57258-P.

References

Buchanan, James, and Gordon Tullock. 1962. *The Calculus of Consent: Logical Foundations of Constitutional Democracy*. Ann Arbor, MI: University of Michigan Press.

Caldwell, Bruce. 2006. Popper and Hayek: Who Influenced Whom? In *Karl Popper: A Centenary Assessment*, ed. Ian Jarvie, Karl Milford, and David Miller, vol. I, 111–124. Aldershot: Ashgate.

Hands, Wade D. 1992. Falsification, Situational Analysis and Scientific Research Programs. In *Post-Popperian Methodology of Economics: Recovering Practice*, ed. Neil De Marchi, 19–53. Dordrecht, Holland: Kluwer.

Jarvie, Ian C. 2001. *The Republic of Science: The Emergence of Popper's Social View of Science, 1935–1945*. Amsterdam: Rodopi.

Lipsey, Richard G. 1963. *Introduction to Positive Economics*. London: Weidenfeld and Nicolson.

Merton, Robert K. 1942. Science and Technology in a Democratic Order. *Journal of Legal and Political Sociology* 1: 115–126.

Morgan, Mary S. 2012. *The World in the Model: How Economists Work and Think*. Cambridge: Cambridge University Press.

Niiniluto, Ilkka. 1998. Verisimilitude: The Third Period. *The British Journal for the Philosophy of Science* 49 (1): 1–29.

Popper, Karl R. 2002 [1934]. *The Logic of Scientific Discovery*. London: Routledge.

———. 1944. The Poverty of Historicism, I, II, III. *Economica*, 11: 86–103, 119–137; 12: 69–89.

———. 1945. *The Open Society and Its Enemies*. London: Routledge and Kegan Paul.

———. 1963. *Conjectures and Refutations*. London: Routledge and Kegan Paul.

Zamora Bonilla, Jesús P. 2002. Scientific Inference and the Pursuit of Fame: A Contractarian Approach. *Philosophy of Science* 69: 300–323.

———. 2008. Methodology and the Constitution of Science: A Game-Theoretic Approach. In *Scientific Competition*, ed. M. Albert et al. Mohr Siebeck: Tübingen.

CHAPTER 5

The Situational Logic of Disciplinary Scholarship

Fred D'Agostino

Jarvie and I

I first became aware of Ian C. Jarvie around 1972. I was then a postgraduate student of anthropology and was writing a lot of essays. Perhaps because I had studied psychology as an undergraduate, I kept trying to cast my essays about cultural and social practices as explanations from the perspective of individuals' beliefs and values. This was sufficiently unusual that my work was characterized as "philosophical." Since I had never studied philosophy, I wasn't sure what that meant: was the comment condemnatory or commendatory? Accordingly, I asked for some help, so that I would know what philosophers thought about and how. Among the things passed on to me were some essays (Wilson 1970), including one Jarvie co-wrote with Joseph Agassi and others by John Watkins and Ernest Gellner. I was lucky to be seconded for a year at the LSE where Jarvie had studied a decade earlier and where both Gellner and Watkins taught, but, once there, I was well and truly hooked and accordingly shifted discipline, university,

F. D'Agostino (✉)
Institute for Advanced Studies in the Humanities,
The University of Queensland, Brisbane, QLD, Australia
e-mail: f.dagostino@uq.edu.au

R. Sassower, N. Laor (eds.), *The Impact of Critical Rationalism*,
https://doi.org/10.1007/978-3-319-90826-7_5

and country of residence in order to pursue what emerged as my calling: to think about human reason and its socially situated and institutionally supported character.

About midway through my career, which is now ending, Jarvie and I conceived the idea of putting together a *festschrift* for Watkins (Jarvie and D'Agostino 1989) on the occasion of his retirement from LSE. As Jarvie wrote, he and I "encountered John Watkins at such different times in his career that [we] have never encountered each other," something which remains true, except when Jarvie got in touch to ask for a chapter on Rational Agency (D'Agostino 2011).

It is thus with considerable pleasure that I offer here some reflections on the situational logic of scholarly publishing, as an academic institution with which both Jarvie and I have been involved (Jarvie 1998): he as long-standing managing editor of *Philosophy of the Social Sciences,* and me, on shorter stints, as editor of the *Australasian Journal of Philosophy* and the journal *PPE: Politics, Philosophy and Economics.* I am particularly interested, as part of a larger project on the disciplines (D'Agostino 2012, 2016), with a puzzling empirical phenomenon: "skewed clustering."

The Explanandum

Like others working through the 1980s–1990s in Anglophone political philosophy, I wrote some articles that took off from or commented on the work of John Rawls, especially his "breakthrough" work *A Theory of Justice* (1971). I was a little late to the party, not entering the fray until 1988, with a paper on Rawls's notion of reflective equilibrium (1988), and it was later still when I became aware that scholarly importance is sometimes measured, by citations, and that my work on Rawls would be measured accordingly. For example, my book *Free Public Reason* (1996) has about 150 citations, whereas that 1988 article has only 12 (Google Scholar 4/25/17).

I was, though I didn't realize it at the time, part of the one discernible "cluster" of publications in political philosophy, as analyzed by Kieran Healy (2013) and which confirms an earlier more general analysis by Phineas Upham and Henry Small (2010, p. 16) that disciplinary research organizes itself (spontaneously as far as one can tell) into exactly these sorts of clusters of co-referring papers and books. No other cluster of scholarly work in political philosophy was visible using Healy's methodology. As a political philosopher, you were either writing about Rawls, or you were "nowhere" in scholarly terms. Indeed, this focus on Rawls's book was a "blockbuster" phenomenon, where "the creation of new knowledge is dominated by a few key insights that challenge the way that people think about an idea" (that's *Theory*), and these sorts of insights "generat[e] high interest and use" (Brouthers et al. 2012, p. 359).

This is the first aspect of my explanandum on this occasion: scholars working in a broad area were, at a certain "granularity," engaged in what Thomas Kuhn (1970) called "normal science"; a "paradigm" or "blockbuster" developed, and the more visible members of the community were engaged in what Kuhn called the articulation of that paradigm. This requires explanation because it is odd that there weren't attractive alternatives to the prevailing Rawlsian orthodoxy. Phillip Pettit's republicanism (1997) and Gerald Cohen's socialism (2009) were later developments.

Also of interest is that of those works that cite *A Theory of Justice* by name, more than 62% have themselves *never been cited at all*, with an additional 12% cited only once. (These are Web of Science Arts and Humanities and Social Sciences citation figures, including the Social Sciences and Humanities Book Citation Index figures, as of 21 April 2017.)

As Brouthers et al. put it (2012, p. 362), describing an entirely general situation: "citation distributions are highly skewed to the right, with a relatively small number of highly cited ... papers, and a relatively large number ... which are cited only once or twice, or not at all." (In Fig. 5.1, the x-axis shows the number of citations the given work has received, and the y-axis shows the number of works with that number of citations.)

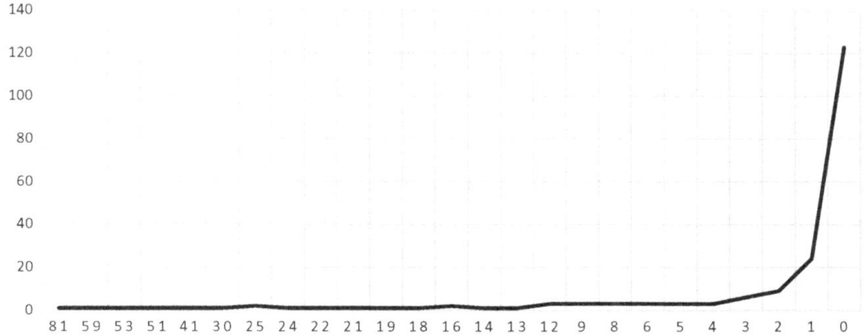

Fig. 5.1 Citations of articles citing *A Theory of Justice*. Source: Web of Science Arts and Humanities and Social Sciences citation figures, including the Social Sciences and Humanities Book Citation Index figures, as of 21 April 2017

This is the second aspect of my explanandum: most work in a cluster is not recognized by citations alone as having contributed significantly to the development of the paradigm by which work in that cluster is informed. This is puzzling, given the widespread assertion that scholars, in their decision-making (What to research? How to do so?), are motivated by their colleagues' recognition, itself a fungible asset (Merton 1968, p. 57), one that can be exchanged for such material complements to the scholarly process as grants, awards, positions, security of tenure, and the like. Richard Whitley's characterization is near-canonical:

> The modern public sciences … constitute a distinct type of work organization and control in which research is oriented to collective goals and purposes through the pursuit of public scientific reputations among a group of colleague-competitors. In such reputational work organizations, the need to acquire positive reputations from a particular group of practitioners is the main means of controlling what tasks are carried out, how they are carried out, and how performance is evaluated … Jobs and resources are allocated largely according to reputations. (1984, p. 25)

Norman Storer concurs that "the institutionally reinforced drive for professional recognition … was the energy that would drive the system" (1973, p. xxiii). This is my explanandum on this occasion. Why do most scholars work on the same problems as other scholars and do so despite the likelihood that they will not be rewarded or recognized for doing so?

The Logic of the Situation

It is useful, as Jarvie (1998), using different terminology, abjures us, to recall what psychologists call "the fundamental attribution error" (Ross 1977). This is where we (not always correctly) default to the assumption that behavior is largely to be explained by an individual agent's personal characteristics rather than by that agent's situation, which may well be shared by others. Indeed, in our particular case, a situational approach is called for independently of guarding against the attribution error, since the explanandum consists of two overlaid patterns formed by the behavior of numerous individuals. An important proviso is where there is no evidence of planning to act in concert (with echoes of Friedrich Hayek 1967).

What we have is a collection of individuals, loosely networked and certainly not constituting a group in any sociological sense, whose independent actions, taken together, create two rather robust phenomena.

First, individuals who haven't antecedently agreed to do so find themselves choosing to work on a problem (out of numerous others). Secondly, they persist in their choices even though unrewarded in the supposed currency of their profession (recognition as expressed through citational practices). Here are some situational factors that might explain these patterns.

Explaining the Clusters

Let's take the convergence on normal discourse (or clustering) phenomenon first. What could explain that? It is important to draw attention to the temporality of the formation of disciplinary communities. Scholars, who themselves have temporally extended careers (D'Agostino 2016) with different opportunities at different stages, become active in these communities at different stages of disciplinary development. For example, early commentary on Rawls was mostly made by scholars who were themselves already reasonably "senior": Thomas Nagel, R. M. Hare, Joel Feinberg, H. L. A. Hart, and Amartya Sen (Daniels 1975), as well as Brian Barry (1975), Robert Paul Wolff (1979), and Phillip Pettit (1974). This amounted to a kind of canonization, and thereafter *Theory* appeared in undergraduate syllabi (and less "senior" faculty had to master the work for teaching purposes). At this point the potential for the work to constitute a paradigm takes shape. Likewise, at this point teaching scholars who master the text think of writing about Rawls's book as a preferred option, relative to other potential topics of enquiry. This is why.

First, writing about Rawls minimizes risk, compared with other options. As Whitley put it, "few are going to risk their futures by trying to publish material which deviates widely from current orthodoxies" (1984, p. 27). Writing about Rawls after 1980 is less risky because there is a receptive audience. It is easier to assess the contribution of submitted works that hew to the orthodoxy than of work which "changes the subject" (or tries to) by introducing new and unfamiliar topics or by approaching even familiar topics in an unfamiliar way (Collins 1998, p. 15). With standard-issue topics and criteria of significance, judgment is easier and more reliable than it is otherwise. Someone who wants competent feedback on their work has a reason, in the situation we're considering, to write about Rawls. Furthermore, if they attend a conference and present a paper, there are sure to be members of the audience equipped to comment with some competence on their work, something that couldn't as reliably be counted on with other choices of topic. (March (1991, p. 73) shows that, other things being equal, feedback about an existing approach is more reliably delivered than for work that explores unfamiliar options.)

Secondly, writing about *Theory* doesn't have the same "set-up costs" (Chubin and Connolly 1982, p. 297) as would be associated with writing about some other author or topic which a scholar hasn't already mastered for teaching or as part of their own graduate training. A scholar who has mastered the Rawlsian apparatus has an investment in that material on which they can continue to draw without significant additional effort. To tackle some other author, text, or problem will typically involve higher costs in this sense. When I first suspected that social choice theory (in the manner of Kenneth Arrow) might be relevant to understanding decision and judgment (D'Agostino 2000), it took me a year to master the apparatus. Since no one else working in the field was familiar with the theory, my use of it in relation to the concept of incommensurability was largely ignored by others working on that topic. (A *Philosopher's Index* search, on 5/3/17 on "Arrow + incommensurability," returned a nil result.)

Thirdly, writing about *Theory* gives an author an opportunity to exploit "positive network externalities" (Teece 1998, p. 523). Because the tools and concepts are the same as those being used by others, they complement each other and draw on others' insights. Writing about Rawls in the 1980s encounters a copious literature which was a "well-trodden path" that represents a materialization of previous work, in Steve Woolgar's terminology (1995, p. 163), so that the individual scholars setting out on such a research "trail" (Chubin and Connolly 1982) unavoidably encounter "off-the-shelf" problems and tools which, as Philip Kitcher put it (2001, p. 112), "stabilize and solidify [and] ... set the scene for the next set" of activities. They have a prefabricated tool kit of concepts, problems, rhetoric, and techniques. They don't themselves have to develop these tools de novo, and, importantly, they don't have to establish the legitimacy of their deployment. Accordingly, such an approach "decrease[s] the cost of learning in the neighborhood" (Zhou 2002, p. 262) and delivers "increasing returns" to participants (Arthur 1994), making it, other things being equal, sensible to adopt such an approach.

We can reasonably expect the contributions of individual scholars to "cluster" around and take the form of commentary on some canonical, paradigmatic, or blockbuster work, such as Rawls's *A Theory of Justice* for reasons that are largely situational. Of course, "individual researchers attempt to choose their research topics 'sensibly'–neither randomly nor omnisciently–in pursuit of bounded self-interest" (Chubin and Connolly

1982, p. 293). Noting that the parameters of "sensible" decision-making about topic selection and other research decisions are subject to considerable uncertainty, they say: "The researcher must decide to pursue a trail with considerable uncertainty as to its final yield; and the uncertainty is greater for studies further down the trail than for the next one or two. It is this rising uncertainty [which means that] ... the researcher's planning horizon is necessarily quite restricted" (Ibid., p. 294). The logic of choice may well be the decision-theoretical calculation of advantage, as Chubin and Connolly suggest. The situation, however, is the primary explanans for the pattern that is formed by individual decisions. Risk reduction, setup costs, and network externalities are all features of the situation of choice, features that are constituted by the disciplinary state of play, including its institutional support, from journal editors to referees. These factors explain the clustering phenomenon. Had they been different, had referees, and had editors, a strong preference for difficult novel materials that challenge current orthodoxies, then individuals, acting individually rather than in concert, would have reached different decisions, yielding a different (perhaps more confused) "pattern."

Explaining Persistence Without Recognition

One of the consequences for most scholars, perhaps an unintended and unexpected consequence, of choosing to write about Rawls was that their articles, notwithstanding that they were "on topic," were not going to be cited by others (see Fig. 5.1). This doesn't mean, of course, that they were not read by others, but, given the conventions of academic publishing, it did mean that others did not regard them as sufficiently compelling to acknowledge in their own writings. This is puzzling relative to a widely influential model of the scholarly community, one which is based on the proposition that individual scholars make decisions about what to write on the basis of expected returns (Bourdieu 1988; Whitley 1984). Indeed, this model has recently been presented in a more sophisticated form by Geoffrey Brennan and Phillip Pettit, in their influential book *The Economy of Esteem* (2004). Why, if citations in some sense measure esteem and esteem is the driver of behavior, do scholars make choices about their work that do not reliably lead to esteem, as measured by citations?

The first thing to observe is that we have to be careful about the explanandum in this case. Most work, no matter the topic, is uncited by others. If we are trying to explain why scholars wrote about Rawls despite the likelihood that their work wouldn't be cited, we have to acknowledge that if they'd written about something else, it probably wouldn't have been cited as well. So the explanandum should be, why do scholars write and seek to publish *at all*, given the likelihood that their work won't be cited and, hence, that the community of scholars will not reward them with esteem? Once we explain that, we can rely on the argumentation of the previous section to explain why political philosophers wrote about Rawls in the 1980s–1990s.

There are a number of relevant factors. First, in the humanities it is important to have published but not as important (at universities where most scholars work) to have been frequently cited by others. When scholars in philosophy come up for tenure, for example, it is their record of publication that is important. It will matter where their work has appeared: How good are the journals? Do they have high rejection rates? Are they prestigious in their specific fields? If there are satisfactory answers to these questions, that will usually be enough to make the case. Except at top-tier universities, how many citations a scholar has accumulated will not be particularly relevant. This is the situation for most scholars, and therefore the fact that they are not discouraged from publishing their work by the likelihood that it will not be cited is intelligible relative to this situation. Notice that this analysis reinforces the analysis of the clustering phenomenon. Certainly, the likelihood of a paper about Rawls being *published* is greater than that for a paper on other topics.

Secondly, it is not clear that the esteem model really works. As I have tried to show elsewhere (D'Agostino 2016), it is arguable, in Jenny Lewis's words, that scholars "are not driven purely by external rewards, but also by intrinsic motivation and social preferences … comply[ing] with social and professional norms, and follow[ing] their own curiosity …" (2013, p. 67). We are expected by our peers to publish our work; we enjoy seeing our work in print; we are caught up, in various roles, in the larger enterprise of academic publication, as referees and as editors. We enjoy the scholarly activity for its own sake, and since writing is an instrument for the clarification of our ideas and for us to assure ourselves and others that we understand the topics we are thinking about, there are, as Lewis says, significant intrinsic motivators for our actions as authors. On this account, scholars are participants, not in a competition for esteem (which almost all of them will "lose") but instead in a practice, in Alasdair MacIntyre's specific sense (1981, p. 190), where their achievements are assessed by journal editors and referees against the standards and practices of the discipline and the era. In this context, political philosophers writing about Rawls could achieve *recognition* as competent members of the scholarly community. And, while recognition doesn't facilitate quite as many opportunities as esteem does, it is fungible in ways that are relevant, indeed highly salient, for most scholars. As Brennan and Pettit point out (2004, p. 218), writing about the work of someone who is highly esteemed may produce a halo effect, so that the secondary literature to which the (uncited) author contributes acquires a kind of borrowed or secondhand esteem. Recognition via "mere" publication will, for many scholars in typical academic situations, be enough to secure employment and a place in the community of scholars.

Again, the matter involves situational factors. It is precisely because important career decisions require a publication (rather than a citation) record that political philosophers can publish without the prospects of thereby acquiring esteem (and will publish on Rawls because doing so is a more reliable route to publication). Writing more than 50 years ago, Storer already enunciated the relevant point: "[I]f the motivational importance of professional recognition lies in its affirmation of successful role-performance, why is it not merited even when an individual has not made an important [i.e. citable] discovery? Simply obeying the rules of the game, contributing to the general well-being of the community through supporting the norms, would seem to be 'successful role-performance'" (1965, p. 57).

As before, when considering the clustering phenomenon, the logic of choice may well be for each individual scholar the decision-theoretical calculation of advantage. When we consider the persistence of "unrewarded" publishing strategies, however, we need to consider two related aspects of the relevant situation of choice. First, for most scholars it is situationally crucial that they be published, even if their work isn't (much) cited. To be published is a "hurdle" requirement for tenure, whereas to be cited is a variable that determines at what university one might hope to be tenured. Secondly, the driving force of scholarly activity (perhaps even for many who are highly esteemed) is not esteem but recognition. And recognition doesn't require citation, merely publication (and other forms of competent participation).

The Social Institutions of Philosophical Publication

To consider the institutions superintending scholarly activity was one of the topics which the editors of this volume suggested to me, as one to which Jarvie had contributed and in which he was interested. As indicated, both Jarvie and I have been editors of journals. Certainly, as an editor, I have wondered and been bewildered about the clustering phenomenon. What an editor can approve for publication is a product of a number of factors, the two most important being what the pool of submissions looks like and what the criteria for publication are. No matter how and where one sets the bar, however, genuine diversity in what appears in the journal depends, in significant measure, on genuine diversity among the papers submitted. And that's what perplexed me as an editor: I just didn't see that diversity, and would have welcomed it if I had. While immersed in what is a consuming activity (largely unrewarded by either esteem or recognition, I might add!), I never managed to understand, let alone do anything effective about, what I have here called "the clustering phenomenon." The opportunity to do something about this has, for me anyway, passed. However, I have some pretensions now, as a result of this very inquiry, to understanding it a bit better and am grateful to the editors of this volume for the opportunity to reflect on this topic and to Jarvie for pointing me, via his work on situational logic, to an appropriate method of engagement. In a situation which creates a "basin of attraction" toward normal scholarship in the Kuhnian sense, the rational behavior of individual scholars creates a skewed clustering in the citational practices of authors.

A puzzle remains. Most work in a given cluster is uncited. We've seen why the work clusters, and we've seen why individual scholars produce work that, predictably, will not be cited. What I haven't explained is why the work *is* uncited. Is this a failure of scholarly professionalism on the part of others working in the field? Haven't they taken the time to discover the writings of others and, a fortiori, cannot quote or cite them? Is a heuristic at work? When I run a search and discover dozens of potentially relevant papers on the topic I'm considering, to control "information costs," I look only at those papers whose authors have high esteem ratings. Certainly, some of my experiences as editor suggest one or both of these explanantia are relevant. Another possibility, however, is that most papers on a given topic amount, in Andrew Abbott's telling phrase (1999, p. 177), to mere "theory churning," or as I would say "cranking the handle," the usual tools are applied to the usual issues with relatively banal results in most cases. Given the situation—publication is a hurdle, but citations are not—many authors will, especially given other pressures on their time, energy, and imagination, produce work that is just good enough to be published (Simon 1956); this is unlikely to be striking enough in its findings or techniques to warrant close attention or the adoption, with appropriate citation, of its claims or approaches. The "non-citers" of uncited work respond rationally to their own situation.

The domain of scholarly publishing, then, is an ecosystem in which the relevant agents respond appropriately, given their own "bounded rationality" (Simon 1991), to the institutionally mediated incentives. A nice illustration, I believe, of the logic of the situation as the appropriate template for the explanation of patterned social behaviors.

References

Abbott, Andrew. 1999. *Department and Discipline: Chicago Sociology at One Hundred*. Chicago: University of Chicago Press.

Arthur, W. Brian. 1994. *Increasing Returns and Path Dependence in the Economy*. Ann Arbor: University of Michigan Press.

Barry, Brian. 1975. The Liberal Theory of Justice: A Critical Examination of the Principal Doctrines in a Theory of Justice. *The Philosophical Review* 84 (4): 598–603.

Bourdieu, Pierre. 1988. *Homo Academicus*. Cambridge: Polity Press.

Brennan, Geoffrey, and Philip Pettit. 2004. *The Economy of Esteem: An Essay on Civil and Political Society*. New York: Oxford University Press.

Brouthers, K.D., R. Mudambi, and D.M. Reeb. 2012. The Blockbuster Hypothesis: Influencing the Boundaries of Knowledge. *Scientometrics* 90 (3): 959–982.

Chubin, Daryl E., and Terence Connolly. 1982. Research Trails and Science Policies. In *Scientific Establishments and Hierarchies*, ed. N. Elias, H. Martins, and R. Whitley. Dordrecht and Boston: D. Reidel Publishing Company.

Cohen, Gerald A. 2009. *Why Not Socialism?* Princeton: Princeton University Press.

Collins, Randall. 1998. *The Sociology of Philosophies: A Global Theory of Intellectual Change*. Cambridge, MA: Harvard University Press.

D'Agostino, Fred B. 1988. Relativism and Reflective Equilibrium. *The Monist* 71 (3): 420–436.

———. 1996. *Free Public Reason*. New York: Oxford University Press.

———. 2000. Incommensurability and Commensuration. *Studies in the History and Philosophy of Science* 32: 429–447.

———. 2011. Rational Agency. In *The Sage Handbook of the Philosophy of Social Sciences*, ed. I.C. Jarvie and J. Zamora-Bonilla, 182–198. Los Angeles: Sage.

———. 2012. Disciplinarity and the Growth of Knowledge. *Social Epistemology* 26 (3–4): 331–350.

———. 2016. Disciplines, the Division of Epistemic Labor, and Agency. In *Social Epistemology and Epistemic Agency*, ed. P. Reider. London: Rowman & Littlefield.

Daniels, Norman, ed. 1975. *Reading Rawls: Critical Studies on 'A Theory of Justice'*. Stanford: Stanford University Press.
Hayek, Friedrich A. 1967. The Theory of Complex Phenomena. In *Studies in Philosophy, Politics and Economics*, ed. F. Hayek. London: Routledge & Kegan Paul.
Healy, Kieran. 2013. *A Co-Citation Network for Philosophy*. https://kieranhealy.org/blog/archives/2013/06/18/a-co-citation-network-for-philosophy/.
Jarvie, Ian C. 1998. Situational Logic and Its Reception. *Philosophy of the Social Sciences* 28 (3): 365–380.
Jarvie, Ian C., and Fred B. D'Agostino, eds. 1989. *Freedom and Rationality: Essays in Honor of John Watkins*. Dordrecht: Kluwer Academic Publishers.
Kitcher, Phillip. 2001. *Science, Truth and Democracy*. Oxford: Oxford University Press.
Kuhn, Thomas S. 1970. *The Structure of Scientific Revolutions*. Chicago: University of Chicago Press.
Lewis, Jenny M. 2013. *Academic Governance: Disciplines and Policy*. New York: Routledge.
MacIntyre, Alasdair. 1981. *After Virtue: A Study in Moral Theory*. London: Duckworth.
March, James G. 1991. Exploration and Exploitation in Organization Learning. *Organization Science* 2: 71–87.
Merton, Robert K. 1968. The Matthew Effect in Science. *Science* 159: 3810.
Pettit, Philip. 1974. A Theory of Justice? *Theory and Decision* 4 (3): 311–324.
———. 1997. *Republicanism*. Oxford: Clarendon Press.
Rawls, John. 1971. *A Theory of Justice*. Oxford: Oxford University Press.
Ross, Lee. 1977. The Intuitive Psychologist and His Shortcomings: Distortions in the Attribution Process. In *Advances in Experimental Social Psychology*, ed. L. Berokowitz, vol. 10. New York: Academic Press.
Simon, Herbert A. 1956. Rational Choice and the Structure of the Environment. *Psychological Review* 63 (2): 129–138.
———. 1991. Bounded Rationality and Organizational Learning. *Organization Science* 2 (1): 125–134.
Storer, Norman W. 1965. *The Social System of Science*. New York: Holt, Rinehart and Winston.
———. 1973. Introduction. In *The Sociology of Science: Theoretical and Empirical Investigations*, ed. N. Storer. Chicago and London: University of Chicago Press.
Teece, David J. 1998. Design Issues for Innovative Firms: Bureaucracy, Incentives and Industrial Structure. In *The Dynamic Firm: The Role of Technology, Strategy, Organization, and Regions*, ed. A. Chandler, P. Hagstrom, and O. Solvell, 134–165. Oxford: Oxford University Press.

Upham, S. Phineas, and Henry Small. 2010. Emerging Research Fronts in Science and Technology: Patterns of New Knowledge Development. *Scientometrics* 83 (1): 15–38.

Whitley, Richard. 1984. *The Intellectual and Social Organization of the Sciences.* Oxford: Clarendon Press.

Wilson, Bryan, ed. 1970. *Rationality.* Oxford: Basil Blackwell.

Wolff, Robert P. 1979. Understanding Rawls: A Reconstruction and Critique of a Theory of Justice. *The Journal of Philosophy* 76 (9): 496–510.

Woolgar, Steve. 1995. Representation, Cognition, and Self: What Hope for an Integration of Psychology and Sociology? In *Ecologies of Knowledge: Work and Politics in Science and Technology*, ed. S.L. Star. Albany: State University of New York Press.

Zhou, Xueguang. 2002. Organization Decision Making as Rule Following. In *Organizational Decision Making*, ed. Zur Shapira. Cambridge: Cambridge University Press.

CHAPTER 6

Academia as Cargo Cult

Steve Fuller

Ian C. Jarvie published my first article, a revised version of an essay that I had submitted for a philosophy competition at Cambridge (Fuller 1983). It reflected a culmination—such that someone in their early 20s could provide—of my early interest in semiotics and French structuralism that was the background noise of my undergraduate years at Columbia and the intense study of philosophical logic and the philosophy of language in which I was engaged as a Master's student at Cambridge. I was pleasantly surprised at the acceptance, and while I never really did anything more with the article (though I might at some point in the future), it opened the door for me to review books by Michel Serres and Patrick Heelan, which informed my original formulation of social epistemology (Fuller 1988, Chaps. 2 and 5). By the time I had completed my PhD at the University of Pittsburgh in 1985, Jarvie was willing to write me a letter of recommendation, which I am sure helped to assuage the concerns of my first employer—the University of Colorado at Boulder—that I might simply be a standard product of what was widely regarded as logical positivism's last stand. As it turns out, my first job involved teaching continental philosophy in a department that was quickly becoming more analytically oriented.

S. Fuller (✉)
Auguste Comte Chair in Social Epistemology at the
Department of Sociology, University of Warwick, Coventry, UK
e-mail: S.W.Fuller@warwick.ac.uk

R. Sassower, N. Laor (eds.), *The Impact of Critical Rationalism*,
https://doi.org/10.1007/978-3-319-90826-7_6

As someone whose entry into philosophy was channelled through the social sciences—still regarded as the low road—I was impressed by Jarvie's ability to leverage anthropology for philosophical purposes. At Columbia, I had been a student of Marvin Harris, the great champion of 'cultural materialism' who taught the introductory anthropology class as a commentary on his magnum opus of the historico-criticism of his field (Harris 1968). If nothing else, Harris conveyed to students the seriousness if not urgency with which we should take the epistemological problems raised by anthropological practice. This training primed me to appreciate that Jarvie's early preoccupation with cargo cults was not simply an idiosyncrasy of either himself or his fellow anthropologists. It pointed to something deeper in our human self-understanding. In the 1970s and 1980s, it appeared as the debate between 'rationalism' and 'relativism', in which Jarvie was a major player. And while the words in quotes remain as labels, they stand for artificial positions which do not really capture what has been at stake all these years. In what follows I take a fresh stab at the matter, starting from Jarvie's original start in cargo cults and moving into our own current practices because 'anthropological' insight—to be worthy of the name—must be able to say something that at once illuminates 'us' and 'them'.

Cargo cults are among the more embarrassing phenomena that anthropologists have had to study, not only because of their inherently odd character but also the unfavourable light in which they put the members of a native society vis-à-vis their foreign interlopers. The phrase 'cargo cult' is normally attributed to the practices of Melanesian islanders who observed and received many modern goods from military forces in successive waves of occupation during the Second World War. However, these unexpected deliveries left the natives in a conflicted state. While they valued the goods highly, the goods prima facie reflected poorly on the natives, who had little of comparable value to trade with the foreigners. The natives tried to justify their psychic plight in various ways, most notably by convincing themselves that the foreigners had stolen the knowledge for making the goods from them, which implied that the natives really could produce the goods themselves. And so, they tried, largely by refashioning what the foreigners had left behind, increasingly embellished with the natives' own self-made attempts to copy the form in which the goods appeared. As for copying the goods' function, the natives resorted to what to foreign eyes looked like 'make-believe'. What struck the young Jarvie in all this was that his fellow anthropologists felt just as compelled as the Melanesians to rationalize a patently irrational response to a situation that posed a strong challenge to native self-understanding.

Jarvie (1963) revealed himself to be more philosopher than anthropologist when he realized that anthropologists displayed the same 'cargo cult' tendencies as the Melanesian natives they studied. In particular, they presumed that there was some rational redemption in store for the natives' prima facie irrational practices. This redemption would usually come in the form of a perceived sense of social stability, the evidence for which lay mainly in those societies having survived as a result of having adhered to those practices over a long period. On this basis, anthropologists characterized the practices as 'functional', thereby spinning them in a positive and perhaps even quasi-intentional direction, as if the practices might be 'unconsciously rational' in some yet to be disclosed way. Starting with Malinowski, Freud proved to be a useful accomplice in this interpretive project, which in turn inspired Robert Merton to distinguish latent (aka unconscious) and manifest (aka conscious) functions.

After all, from the anthropologists' own considered standpoint, the practices were indeed prima facie irrational, if we take the natives' own accounts of those practices at face value. For example, a spell at most induces the placebo effect in its recipient, who is now primed to interpret any signs of comfort or discomfort as evidence of the spell working its magic. More likely, the spell operates performatively to licence those who witness or hear about the spell to treat its recipient in a significantly different way. Perhaps the combined effect of all these interpersonal reorientations is one of social equilibrium. A more cynical observer might regard them as constituting an elaborate face-saving ritual. In any case, the spell does not do what the natives think it does.

Moreover, even if the natives—or at least some of their members—realized that the spell did little more than secure social order, it would still not suffice to make the practice rational. In that case, one would need to weigh the costs and benefits of failing to proceed by some other, non-spell-based means, even granting the legitimacy of the natives' own social order. This would involve the assessment of what economists call 'opportunity costs'. Of course, some of the possible alternatives to casting a spell, such as modern medicine, may not be readily available to the natives. But then the anthropologist would need to consider the costs and benefits of introducing them. In that respect, the anthropologist and the natives are always co-implicated in a sense of rationality that is *at the same time* (pace Habermas) communicative and instrumental. Interestingly, even Marvin Harris, the avowed 'cultural materialist' anthropologist who understood native practices as contributing to a society's 'mode of production', tended to regard long-standing societal norms, such as the Hindu prohibition on meat-eating, as implicitly optimizing over the Indian ecology—on the basis of relatively little economic analysis of alternative courses of actions. He, too, settled for a version of 'functionalism', but one in which Marx replaced Freud. In more recent times, sociobiologists and evolutionary psychologists keen to uphold the value attached to the sheer long-term survival of practices have replaced Marx with Darwin to similar effect.

Whatever the outcome of opportunity cost calculations, a couple of things are clear. First, it should not be taken for granted that the ongoing perceived stability of a society is indicative of its rationality. One first needs to determine alternative courses of action, why they have not been chosen and what would have been the most likely consequences had they been chosen. Second, insofar as the anthropologist can interact with the natives sufficiently well to be able to make sense of their customs and beliefs in some detail, they share a common notion of rationality—at least common enough for the anthropologist potentially to offer criticism of the rationality of the natives' practices. However, as Jarvie realized early on, it is precisely the prospect of criticizing the native interlocutor that the anthropologist appears to want to avoid at all cost: to wit, that the natives could simply be wrong on their own terms and that their practices ought to be changed.

The superstitious character of cargo cults appears in the anthropologist's behaviour at two levels: in terms of the perceived redemptive power of the natives and the imagined meta-redemptive power of the anthropologist. Taken together, they amount to presuming both that the natives sort things out in the best possible fashion when left to their own devices and that the anthropologist's imagined radical detachment from the native society puts him/her in a particularly good position to recognize that fact.

After all, the ease with which the anthropologist puts a positive spin on the natives' beliefs simply based on the continuity of their collective existence is probably something that s/he would probably not extend to the beliefs that hold together the anthropologist's own society. In other words, the anthropologist would normally be critical, even suspicious, of the surface stability of the society s/he normally inhabits as it likely masked hidden tensions and conflicts of one sort or other, perhaps even 'structural contradictions', as Marxists would say. However, by presuming stability on the part of the native society and instability on the part of the anthropologist's own society, s/he effectively re-enacts Europe's own trauma over the passage from 'tradition' to 'modernity', which provided the crucible for the formation of the social sciences. It is as if the natives formally studied by the anthropologist function as an object of nostalgia for an imagined premodern European world, which was much more harmonious than the current one.

Moreover, this vision of native harmony is masked by a complementary self-understanding on the part of the anthropologist, namely, that s/he is a Martian-like 'objective' observer, who is so detached from the natives, that s/he can judge them rational without having to claim that s/he shares anything much with them with regard to norms. This image of the anthropologist as Martian, which became very widespread in analytic philosophy, was due to Arne Naess, whose doctoral dissertation, done under aegis of the Vienna Circle, focussed on how such a being would go about trying to communicate with Earthlings (Naess 1965). However, Jarvie realized that if we imagine the task in this way, what W.V.O. Quine—like Ness, a junior colleague at the Vienna Circle—would later popularize as a 'radical translation' episode, then it is not at all clear that the hypothetical Martians would be able to distinguish humans from other earthly life forms or indeed anything else on our planet that the visitors happened to find semiotically rich (Jarvie 1986, p. 186).

In effect, the Martians could operate with a radically naturalistic conception of the Earthlings in the spirit of, say, a Darwinist, for whom natural selection operates as an equal opportunity *unemployer* across all species insofar as, on Darwin's view, the blind character of 'selection' is really about 'deselection'. This may indeed what Naess and (more likely) Quine intended, but such an interpretive stance does not constitute 'rationality', strictly speaking, because it applies asymmetrically, as it were 'vertically' rather than 'horizontally' between the Martian anthropologist and the native Earthlings. It is 'non-communicative', from a Habermasian standpoint. In other words, the workings of natural selection—known only to the Martian, in this case—are presumed to be true in a way that is incontrovertible by whatever arguments or claims that the Earthlings might make against it. That the Martians might propose this hypothesis successfully without its being seen as a threat to the Earthlings is simply a testimony to the Martian mastery of earthly etiquette—given that the hypothesis is true in any case.

The point that I take to lie behind Jarvie's original critique of cargo cults is that the very claim to find 'rationality' of a radically different kind from our own amounts to a conceptually incoherent form of escapism on the part of the anthropologist, who does not want to see his/her own and the natives' society as part of a common world in which one, the other, or both could get things sufficiently wrong that someone else could envisage an alternative, better course of action for them. Thus, the Martian metaphor is appealing because it allows the anthropologist to maximize the distance between one's own and the other's societies, making it difficult to apply any sort of logic to the situation.

After all, understood as a social practice, logic presumes that if it is possible to establish what people believe, and especially to understand those beliefs as constituting a coherent system, then it is possible to hold those people accountable for their beliefs even if we don't share most of them. In that respect, the ultimate mark of a true understanding of what the natives believe—as opposed to a simulacrum of their beliefs that simply caters to the anthropologist's own society—is that the anthropologist could persuade them that it is in their own interest to conduct themselves differently. Anything short of that, as I have argued, potentially renders the anthropologist guilty of wishful thinking with regard to the native society and hypocrisy with regard to his/her own society.

If the reader still finds something strange about this argument, it is because it raises Popper's critical rationalism to a categorical imperative for human conduct, one that would bind together the anthropologist and the natives in a common normative order, notwithstanding the clear analytic difference between the societies of the anthropologist and the natives. However, as Jarvie's early work showed, anthropologists have been historically in search of an illusory middle ground between complete disengagement from the natives under study, whereby the anthropologist would effectively leave them alone as objects of naturalistic observation, and the Popperian position, which involves full engagement with the natives, including the prospect that the anthropologist might criticize them and vice versa.

To be sure, any serious cross-cultural encounter requires a measure of tact and tolerance on both sides. However, anthropologists turn into epistemic cargo cultists when they treat what are in effect the means they use to gain the confidence of the natives into the ends of anthropological inquiry itself. In that case, anthropology simply becomes an account of how to fit in—or 'pass', as the ethnomethodologists would put it—among members of the native society, but with a twist. The twist is that the account is designed not for the natives but the anthropologist's own society, as if the anthropologists need to convince themselves that they are dealing with the natives in the right way. Thus, any account of the natives becomes an exercise in what economists call 'virtue signalling'. But doesn't it just end up being an exercise in misplaced flattery that neither tests the Martian's claims nor benefits the natives on whom they so easily pass positive judgement?

In the process, anthropology is reduced to a set of rhetorical moves designed to demonstrate that the anthropologist had not made any significant errors or enemies among the natives. But that standard falls far short of a 'deep' (i.e., causally relevant) understanding of native society, which would require that the anthropologist demonstrate the capacity to offer the sort of advice if not outright criticism that the natives would regard as an improvement on their own terms. Only then would the anthropologist show that s/he has grasped something that truly matters to the natives. The same standard could apply vice versa, with the natives offering the anthropologist advice that could be brought back to his/her own society. To be sure, more 'politically correct' anthropologists and their boosters, such as Donna Haraway (2016), are full of lessons that the natives might offer to modern societies, especially with regard to ecological sustainability. But tellingly, the advice flows only one way. These same people do not countenance our telling the natives that they, too, should change their ways.

This illusory middle epistemic ground in which anthropology has historically dwelled—the source of its own cargo cult mentality—has expanded into an *academic expressivism* that is exemplified by the rise of so-called identity politics. I say 'so-called' because 'identity' needs to be understood in a particular way, namely, as referring to the performance of a specific group identity, which 50 years ago might have been pejoratively called a 'stereotype'. Academics take it upon themselves to demonstrate in their own writings, speech, and self-presentation the identity of a community that is in some sense 'marginalized' and hence not normally given the representation it deserves. This leads academics to channel their understanding of the group through some basic disadvantages that it has vis-à-vis more dominant groups. And while this understanding serves to stabilize the group's identity for analytic purposes, it also amounts to a political holding pattern.

Because, say, African-American identity has been formed under the regime of 'white privilege' as the expelled 'other', it is difficult for identity politics academics to conceptualize what African-American identity would look like without such a regime in place. Their failure of imagination is then explained—and effectively rationalized—in terms of the absolute hold that white privilege has over the situation. Nevertheless, academics believe that if they declare this racial hegemony long and hard enough, something miraculous will happen to overcome this political impasse. If nothing else, identity politics displays enormous faith in the power of 'witnessing' the oppression of others. Indeed, in perhaps the most influential presidential address to the American Sociological Association in living memory, Michael Burawoy (2005) called for a 'public sociology', which I subsequently derided as 'car park sociology' because it would gut the intellectual tradition of the discipline on behalf of providing a platform for self-identified groups to seek public advocacy (Fuller 2016). In effect, those groups could 'park' their grievances with sociology, which would give them an indefinite airing without necessarily promising any resolution. If anything counts as 'cargo cult on demand', this is it.

Before briefly saying how I think the genuine issues that inform identity politics should be addressed, let me catalogue the three forms that such academic expressivism has taken in our times:

1. Perhaps most obvious and annoying is the repeated invocation of the jargon of one or more living or dead social theorist, which aims both to conjure up their authority and 'call out' the status quo. This is presented as 'critique' in its purest form—namely, speaking truth to power. Unfortunately, 'power' does not seem to recognize this mode of address. Academics appear to think that if they continue to emphasize the world-historic significance of some pronouncement by, say, Hannah Arendt or Judith Butler to rapturous applause at their conferences, it will increase the chances that change along desirable lines will occur. Yet the unlikelihood of that happening does not really matter because academics are reinforced by their system's reward structure to carry on as they have, just as long as they continue to recognize each other's contributions through citations and other performances of mutual esteem. This, then, licences periodic public protest by academics to demonstrate 'solidarity' in the face of potential disruptions to their neat system. Case in point is an event happening as I write this piece, namely, the largest academic strike in UK history. The strike is about the long-term sustainability of a pension scheme that academics probably never took the time to fully understand when they signed up to it, as that would undoubtedly have posed a distraction to their ongoing scholarly campaigns

against neo-liberalism, a form of virtue signalling that also provides them with research credit. I leave it to the reader to disentangle the various levels of irony in the situation.

2. The blind mirroring of the alleged oppressor in one's 'anti-oppressive' strategy is a common trope. What currently passes for 'postcolonialism' amounts to inverted colonialism. Considering the historic importance of China to the world-system, it is striking that postcolonialism's aspirations to provide a truly 'global' understanding of humanity extend no further than the parts of the world that Europe colonized. To be sure, this point had been clear in postcolonialism's founding text, in which Islam was portrayed as the 'East', the de facto 'other' of the 'West', understood as secular Christendom (Said 1978). However, today's postcolonialists have seemingly forgotten this point—hence the conspicuous absence of any discussion of China. A truly global postcolonialism might have pushed back from Edward Said's characterization of the East/West binary, since Said's point was based on the rivalry between Christianity and Islam as two offspring of the same intellectual roots with comparable aspirations. But instead, postcolonial scholars have simply come to occupy the same positions, such that in their 'projective' style and 'touristic' tone, Said's old nemesis, the Princeton Orientalist Bernard Lewis, easily morphs into that of the darling of Indian subaltern studies, the University of Chicago's Dipesh Chakrabarty. The difference is that whereas Lewis focused his Occidental gaze on a floundering Orient, Chakrabarty simply reverses the gaze by turning the telescope around to observe Occidental floundering from an Oriental standpoint.
3. Perhaps the most intellectually interesting and potentially destructive form of academic expressivism involves the curious repurposing of Robert Nozick's (1974) entitlement theory of justice into a theory of restorative or 'reparative' justice. The guiding intuition here is that group disadvantage is ultimately a matter of some prior act of dispossession, and justice consists in restoring the illegally acquired property to its rightful owners. One may understand this policy either quite literally in terms of repatriating land that was colonized by foreigners or somewhat more figuratively in terms of providing equivalent compensation to the descendants of those who had been dispossessed. With the exception of cases involving the lands of indigenous people whose relative isolation from more dominant populations makes them easy to identify, most claims to reparative justice are of the latter variety. Today's summary dismissal of the

original objections to Nozick's entitlement theory is breathtaking. Those objections are presented as simply reflecting the liberal establishment's failure to recognize the harms inherited by society's disadvantaged groups (e.g., Forrester 2018). But in fact, the original objections centred on an empirical and a normative concern. Empirically, it is difficult to establish who should make what transfers to whom, given the historical nature of the injustices and the dynamic and intertwined character of subsequent generations. Normatively, it is by no means clear that those whose lives continue to be marked by past harms are best served by simply undoing the harms, given the world in which they actually live. Note that neither of these concerns involves denying that significant group harms, including dispossession, have been committed in the past. What is denied, however, is that fixating on that fact is likely to do much good for the intended beneficiaries. But here's the rub: the people pushing to revive the entitlement theory of justice are humanists who are most comfortable with the sort of historical research that the theory favours. Thus, much energy is expended on documenting the atrocities of the past to provide rhetorical ballast for claims to reparative justice, as if it were obvious that greater knowledge of the past is a reliable guide to future action in these matters. On the contrary, what should be truly obvious is that the primary targets of just policies are people living in the present and the future, not in the past.

When reflecting on the above three forms of academic expressivism, I am reminded of Abraham Maslow's riposte to Nietzsche's injunction that one should philosophize with a hammer: If all you have is a hammer, everything looks like a nail. Like the Melanesian cargo cultists and their anthropological observers, academics more generally are inclined to think that whatever happens to be at their disposal is likely to be efficacious in setting the world right—what cognitive psychologists dub the 'availability heuristic'. The physicist Richard Feynman had realized this point nearly a half century ago, while delivering the 1974 commencement address at Cal Tech. He coined the phrase 'cargo cult science' to characterize academic research that reproduces the trappings of science without possessing the external validity of science (Feynman 1985, Part 5).

The problem here is at a metalevel to the one played out in the 'science wars' of the 1990s. The problem then was that humanists allegedly did not understand the science on which they mounted their claims. The problem now is that what humanists do understand may not be appropriate to achieve their stated goals, however competent they may be at, say, recounting matters of oppression in the past. Indeed, it may be that the world has changed, while the humanists were not watching. After all, the triad of race, class, and gender became ensconced as sociological categories in the nineteenth and twentieth centuries as a reflection of how states organized their populations in periodic census exercises. However, with the rise of fluid self-definition across all these categories, the 'objective' foundations of the normative order established on the basis of race, class, and gender have been effectively undermined, if not delegitimated entirely. Moreover, this is not simply, or perhaps even primarily, due to some imagined 'postmodern' turn in the social sciences. Equally corrosive of the sociological triad have been the findings from a host of natural sciences on the borderlands of the social sciences—from human genomics to cognitive neuroscience—which challenge the criteria by which one comes to be identified of, say, a particular race or gender.

Not surprisingly, then, the state and sociology are both suffering identity crises in our own time. The concept of 'intersectionality' that is popular among academic expressivists to capture this state of flux is more a symptom than a solution to the problem. In the end, the fate of sociology's categorical holy trinity and the fate of those who would be subsumed under those categories need to be kept separate, with the latter given priority over the former. This point should be obvious to anyone who is not an academic expressivist. However, the academic expressivist clings to race, class, and gender as a cargo cult that provides an illusory sense of control over our understanding of the social world. Indeed, unlike 50 years ago, one nowadays rarely encounters research that is designed to calibrate sociological categories against the self-understandings of the people who would be covered under them. It may well be that the fluidity of people's self-definition of race, class, and/or gender simply reflects the relative lack of importance that they place on the categories, preferring instead to address their problems as individuals or as part of self-organizing groups, the identities of which academic expressivists are unlikely to grasp as long as they persist in their faith in the holy trinity.

References

Burawoy, Michael. 2005. For Public Sociology. *American Sociological Review* 70: 4–28.

Feynman, Richard. 1985. *Surely You're Joking, Mr Feynman!* New York: Norton.

Forrester, Katrina. 2018. Reparations, History and the Origins of Global Justice. In *Empire, Race and Global Justice*, ed. D. Bell. Cambridge: Cambridge University Press.

Fuller, Steve. 1983. The 'Reductio ad Symbolum' and the Possibility of a 'Linguistic Object'. *Philosophy of the Social Sciences* 13: 129–156.

———. 1988. *Social Epistemology*. Bloomington, IN: Indiana University Press.

———. 2016. Beyond Car Park Sociology: Recovering the Roots of Public Sociology. https://www.thesociologicalreview.com/blog/beyond-car-park-sociology-recovering-the-roots-of-public-sociology.html (16 May).

Haraway, Donna. 2016. *Staying with the Trouble: Making Kin in the Chthulucene*. Durham, NC: Duke University Press.

Harris, Marvin. 1968. *The Rise of Anthropological Theory*. New York: Thomas Crowell.

Jarvie, Ian. 1963. Theories of Cargo Cults: A Critical Analysis. *Oceania* 34: 1–31, 108–136.

———. 1986. *Thinking About Society: Theory and Practice*. Dordrecht, NL: Kluwer.

Naess, Arne. 1965. Science as Behavior. In *Scientific Psychology*, ed. B. Wolman and E. Nagel, 50–67. New York: Basic Books.

Nozick, Robert. 1974. *Anarchy, State and Utopia*. New York: Basic Books.

Said, Edward. 1978. *Orientalism*. New York: Random House.

CHAPTER 7

Our Current Sense of Anxiety

John A. Hall

Introduction

I first met Ian C. Jarvie in the London School of Economics in 1972. I was a young graduate student, deeply influenced by Karl Popper, attending his small seminar on rationality and relativism. For some silly reason I argued with Ian, trying to stress the limits of reason. He won the argument hands down, but I remain grateful for that session, for the pleasures of working with him, and for all that he has given to us. Still I return here, a little ungraciously, to the limits of reason, hoping that he will forgive me for so doing given the current sense of anxiety that pervades advanced liberal society. The two great intellectual pillars of his life have been Popper and Ernest Gellner, and I consider their work as a means to locating the difficulties that now face us.

Sociology and Philosophy

My own *maître a penser* has been Gellner, and I approach Popper essentially through his eyes. Gellner often noted that the greatest influence on his own work was *The Open Society and Its Enemies*, both because of its

J. A. Hall (✉)
James McGill Professor of Sociology, McGill University,
Montreal, QC, Canada
e-mail: john.a.hall@mcgill.ca

R. Sassower, N. Laor (eds.), *The Impact of Critical Rationalism*,
https://doi.org/10.1007/978-3-319-90826-7_7

defense of liberalism and its concern with science (see Hall 2010 for a full account). But this did lead him to accept Popper's work as a whole. There were of course differences of temperament that derived from their backgrounds in Prague and Vienna, but I leave these to one side. A deeper personal difference concerned the need for certainty. Popper was a leading exemplar, as were Judith Shklar and Isaiah Berlin, of a particular form of postwar liberalism that insisted that all historicist thinking was dangerous. What mattered for such thinkers was rigorous clarity, the puncturing of balloons purporting some form of necessity. Such thinkers are wholly admirable and one tangled with them at great risk, for their arguments tended to be razor-sharp. Nonetheless, Gellner chose to do so. I suspect that he always felt that clarity in and of itself was not a firm enough basis on which to ground one's values. Certainly he always regretted Popper's move away from social philosophy to almost complete concentration on the philosophy of social science. Just as importantly, Gellner had added social science skills to his philosophical training, and this made him feel that our social world should and could be defended for sociological reasons. Social considerations existed to ground our faith in liberal society. There is a refreshingly philistine quality to Gellner's work at this point, a sort of humdrum materialism, and his position was designed to appeal widely beyond circles of intellectuals. His arguments made their first appearance in 1964 in *Thought and Change*. The basic consideration was that of a very limited form of historicism, a severely truncated form of social evolution: a modern social contract existed in which legitimacy should and would be given to social orders that provided wealth and in which rule was exercised by those co-cultural with the majority of the population (Gellner 1964, Chaps. 1 and 2). Popper admired the book for many reasons but essentially refused to accept its argument. He disliked even this mild historicism, rejecting Gellner's insistence that thought should be influenced by social change (Hall 2010, pp. 158–159). The greatest source of change *was* thought, in Popper's view, thereby defending a strikingly intellectualist position for which Gellner had little sympathy.

Gellner continued to assess Popper's work for the rest of his life, always seeking striking formulations to justify the position that he had established at the start of his career. He was prompted to do so by what he considered to be Popper's increasing insistence on a quasi-Darwinian view in which the growth of knowledge was written into the nature of human life. Two related arguments deserve to be remembered, both of which have considerable merit.

Gellner had a keen sense of the power of ideology, both from personal experience of classic modern ideologies such as psychoanalysis and Marxism and from fieldwork among the Berber tribesmen of the High Atlas as well, of course, from his disciplinary background in social anthropology. The central point that he made again and again was that such ideologies were often circular and self-reinforcing. Differently put, closed systems of social thought were so to speak coherent and solid. One could not escape them. But escape *had* taken place, with the power of Western science being seen by Gellner as one of the key elements of the modern age, above all, as the progenitor of wealth. He spent a great deal of time trying to rationally reconstruct how it had been possible to exit circular worlds, stressing—in a surprisingly intellectualist manner—the important role played by codification within world religions (Gellner 1988, Chaps. 2–5). But he added to his account of the origins of this new world an account of how it worked. What mattered was the establishment of a norm of judgment that had the capacity to humble circular systems. Positivism—less as Comte's religion of humanity than as basic empiricism—has a background assumption of atomism, a belief that experiences are separable, that allows this to take place.

> [Empiricism] enjoins, above all, a sensitivity to the distinction between that which is publicly verifiable by experiential evidence and that which is not, a sensitivity which is low or even systematically obscured in traditional culture. It creates a distinction where it barely existed .... Thus the really important social impact of the ghost philosophy [empiricism] is the injunction "Be sensitive to the boundary, and impose consistency with respect to it"; the secondary injunction, "Down with the transcendent" (*Burn the books containing it* according to Hume, or *Call it technical nonsense* according to Ayer) does not matter much. If the first injunction is well observed and implemented, for all practical purposes the second is already performed and prejudged. In social contexts in which the first is well diffused and respected, it is perfectly possible to play down and, at a superficial level, ignore the second in the interests of courtesy, kindness, tact or antiquarianism. It hardly matters. (Gellner 1974, p. 122)

In this matter Gellner is supremely sophisticated. It may not be the case in reality in itself that sensations are separable. To the contrary, it may be that the world is made up of glutinous mass. Nonetheless, pretending that the world is atomized—that experiences come one at a time—is a background assumption that seems to have worked, with Gellner being prepared to say that the assumption might then be true, although we cannot prove that to be the case. Differently put, empiricism/positivism is an ethic of cognition, a normative commitment with enormously important consequences. So, positivism is a social world, but one that seems to work—or, in his formulation, positivism is right for Hegelian reasons (Gellner 1985). To make this argument was of course to argue against Popper. Mere criticism would not lead to the growth of knowledge; only criticism backed by the sharp questioning within a world that accepted the norm of empiricism would lead to intellectual advance.

The second general argument is an extension of the first. In a brilliant passage, Gellner sought to clarify the positions of Thomas Kuhn and Popper, especially given Popper's confusing insistence that real science was revolutionary but decent politics best served by piecemeal social engineering. To begin with, Gellner suggested that Kuhn was entirely correct to locate a paradigm shift, namely, that between traditional and modern thought, but wrong to underplay the amount of change made possible by empiricism within the latter. With that in mind, he chided Popper both for failing to realize that *the* revolution in science is that of the introduction of empiricism. Scientific change and social development are then both essentially piecemeal within that world (Gellner 1974, pp. 168–184).

Gellner placed great hopes upon his *Legitimation of Belief*, insisting that his general position was spelled out there in its most complete form; hence, we must follow his argument, so as to move from critique of others to the statement of his own view. The overlap is in fact considerable. When discussing modern science, he expands what has been already noted in regard to empiricism in two ways. First, he notes the weakness of empiricism as a model of the mind, while continuing to insist on its enormous merit as a tool by means of which to judge theories. This leads to the second point: science depends as much upon what he terms mechanism as it does upon empiricism. The construction of reductive, elegant, and parsimonious models of "structures" is totally necessary, with empiricism being best designed, as noted, to sort out which of these models seems to work best.

Gellner ends his treatise with reference to four planks, making up a more or less serviceable raft for us to survive in the modern world. This is his grounding of advanced liberal society. Two of these planks concern the basis of modern science: empiricism and mechanism are the cognitive ethics at the heart of the epistemology that explains the extraordinary growth of knowledge in the modern world. But he added to this two much more sociological considerations, which return to his earlier view of the modern social contract. In the years following *Thought and Change*, Gellner had amplified time and again the importance of economic growth. Modern liberal capitalist societies were based on a central contradiction: social inequality was present but those at the bottom had rights. The system worked as long as there was sufficient money both to buy off discontent and to ensure that the unequal slice of the collective pie grew each year (Gellner 1979). The message of this world, he often wryly noted, was less "workers of the world unite" than "consumers of the world unite." Finally, Gellner had begun to add to his simple statement that legitimacy depended upon the leaders of a society being co-cultural with its members. The basic equation had to be met for sure, but what mattered as well was the spirit in which it was held. Nationalism needed to be present, for some greater sense of belonging was called for in the interests of daily understandings, allowing a basic sense of social solidarity. But nationalism should not be taken too seriously; certainly it should never create a world closed unto itself. Hence, he recommended a spirit of "ironic cultural nationalism," something likely to be possible in a world of economic growth in which, he might well have added, massive ethnic cleansing had already taken place.

I admit that I accepted a very great deal of Gellner's general position, liking its dull materialism and cold rigor, while remaining a little worried that there was insufficient room for political liberty within his design. More particularly, Gellner's mild confidence and certain hope that political liberalization would necessarily follow as societies sought to allow room for the technical knowledge that would ensure social stability seemed to me to be overstated, lacking a sense of political conflict (Hall 2010, pp. 191–206). Nonetheless, I felt extremely close to his basic position. Now I am not so sure. I feel anxieties in the most visceral manner for sure, but believe they haunt many in the contemporary world.

Reason Assailed, Inequality Unloosed, Nationalism with Teeth

One has the feeling these days that history is, as Arnold Toynbee had it, "on the move again." Postwar certainties, derived from the lessons taken from the era of two world wars, seem to have dissolved. The argument that follows is simply to say that we are accordingly all at sea, necessarily filled with anxiety. Three particular arguments need to be considered.

Let us start with the horrible and notorious claim, made by Kellyanne Conway, the confidante of Donald Trump, that there are facts and "alternative" facts. Perhaps some comfort can be found in the fact that in many quarters the claim was laughed out of court. Furthermore, it has been wonderful to see people on the streets demonstrating for science: protests *for* positivism indeed! But any high level of confidence is not justified. For one thing, we should remember that there always have been different social worlds within advanced societies. The *uses and gratifications* school of media studies, for instance, has demonstrated that reality is differently perceived. Such differences of perception often occur within a world where facts do have some ultimate hegemony. But this has not always been the case. On rare occasions in the postwar world one has felt oneself to be in Max Weber's world, of different gods seeking different goals. But what was unusual is now becoming normal. One absolutely crucial point to be made here concerns changing technology. Cass Sunstein (2017) describes all sorts of research, from behavioral economics to standard sociological research, that shows the increasing extent to which modern social media cage people within intellectual silos. This can be the result of choice, but it is also something that is done to us, both by providers, such as Facebook, and by political entrepreneurs, keen to use the latest technology to cement identities. *Facts* rule less and less in this new social world, with a good deal of the evidence that Sunstein cites showing that life in such silos does tend to harden political attitudes. The social heterogeneity that gives us the chance to challenge and to change our views is being lost; homogeneity rules within the intellectual silos that are gaining greater and greater salience. Two particular events made this clear. The campaigners for Brexit made claims about the costs of belonging to the European Union that were proven to be untrue. But the Brexiteers did not withdraw the claims, repeating them endlessly instead, secure in the knowledge that those they wished to reach would not have seen any rebuttal. Trump's campaign exhibits the same tendency, albeit to a level hitherto unimaginable. In consequence, we now have a new language of facts and "nonfacts" or "false facts." This development is so severe and dramatic that it undermines the views of Popper as well as Gellner. Bluntly, it takes one's breath away. But there is more, at least if we are to examine the planks in Gellner's raft.

The plank that imagined social peace resulting from ever-increasing wealth is under assault as well. In fact, a little more than wealth was always involved. The disasters of the middle of the twentieth century led to a historic class compromise in the advanced West, the removal of left and right extremes with a promise of full employment often backed by Keynesian policies. It may well be that what mattered most in the postwar period was something slightly different. Walter Scheidel's superb *The Great Leveler* (2017) has demonstrated conclusively that inequality always has a tendency to increase, interrupted only by violent episodes, above all those of war and revolution. So, Gellner's plank was very fragile: it rested on a historical conjuncture—that of the leveling of incomes caused by total war based on mass conscription—and not on some general social evolutionary dynamic. None of this is to say that there is no growth in the advanced world. There most certainly is, albeit with more in the United States than in Japan or Western Europe. But the change is very great. It is no longer the case that one's piece of the pie, however small, will increase year by year. The gains from growth have increased the incomes of those at the top, while exclusions are now massive. Further, it may well be that the technological revolution of our time will not produce high levels of employment, as had been the case in previous historical periods. Neoliberalism and austerity rule over the poor, and a very great deal of resentment did become apparent during the financial crisis that began in 2007.

The final plank in Gellner's raft is not so much rotten as increasingly absent. In this context we should remember that interwar Vienna produced masterpieces not just from Friedrich Hayek and Popper but from Karl Polanyi as well. Polanyi's *The Great Transformation* (1944) can be considered in contemporary circumstances as the single most important text in social theory that helps us understand our condition. The treatise seeks to explain the disasters of the twentieth century by means of a single phrase. Capitalism demands endless change, disrupting all and everything to such an extent that sooner or later *society will seek to protect itself*. That phrase helps us understand the previous point, but we need to add into this formulation (more clearly than did Polanyi himself) noneconomic factors. What matters more than change is the sensation that it has been caused by outsiders of one sort or another. If it was Jews in the German case, it was Islamists for Trump and both migrants and continental workers in that of Brexit. There are of course all sorts of extra elements at work in each of these cases. For example, Brexit would not have been possible without the ruling elite still suffering from imperial delusions of grandeur, still failing, as Dean Acheson cruelly but accurately had it decades ago, from having lost an empire and failed to find a role. The result is absolutely clear. There is nothing ironic about the way in which nationalism has been embraced in key segments of the United States and Great Britain, and to a lesser extent in most countries of the advanced world. Real viciousness is involved here. In my own Quebec, the Parti Quebecois' playing of a xenophobic card seems to be directly related to a rise in hate crimes, part of a much more general pattern. Lives are being lost. (I rely here on research by Matthew Lange, McGill University.)

What Is to Be Done?

Gellner's case against Popper was that he was naïve. With great regret—for it hurts me too—my argument has been that his position now looks slightly complacent. Times have changed and anxieties abound. Perhaps one should not despair. Trump lost the popular vote, and the margin between Remainers and Brexiteers was not large. The rise in social inequality in social democracies is not yet great. Further, the European Union may pull together in the face of Brexit so as to deal with some of its looming problems, not least as it seems as if the challenge from the right in France in 2017 was defeated. Still, the Trump presidency and Brexit have had and will have great consequences, so it behooves us to ask what can be done to deepen support for liberal democracy.

It makes sense to begin with the occasional and ad hoc proposals of Sunstein (2017, Chap. 9) that follow his analysis of the dangers posed by a divided democracy. These include the creation of deliberative domains, disclosure of the political links of producers of communications, voluntary self-regulation, subsidies to encourage more challenging media performance, "must-carry" policies designed to increase public awareness, and the use of links to draw attention to alternative sources of information. This is a ragbag, open to the objection that they can only become institutionalized once one has power. But how does one gain power? Protests for positivism help, of course, as does struggle, more generally. But so, too, does really serious critical analysis, especially for intellectuals who lack the efficacy on barricades that they sometimes have in the library. To say this is to return us to Popper's claims, albeit without the confidence that undergirds his position. It is also to remind us how very much Gellner is missed: no postwar thinker was better at acidly dissecting dreams and illusions of one sort or another.

If that is a call to action, it is not designed to say that the forces discussed can be ignored or easily defeated. There should be no sudden move from a moment of pessimism to ungrounded optimism. Any re-grounding of confidence in liberal democracy is likely only to come as the result of severe setbacks. One can see this in the case of climate change. Agreements have been made, but they are not being enforced. Only some catastrophe—say the move of a massive number of people consequent on the flooding of a country, such as Bangladesh—will bring action. The same is true politically. The resentment that will eventually be unleashed against the Trump presidency and the Brexiteers is not something that one looks forward to, especially as the process involved is likely to be dangerous and deeply unpleasant. But both will teach a lesson. Kant told us that the presence of war was necessary as it reminded us of the importance of peace. Human memory is short, and the lessons of the two world wars have been forgotten. Let us hope that we can remember for a longer period the lessons we are about to learn.

References

Gellner, Ernest A. 1964. *Thought and Change*. London: Weidenfeld and Nicolson.

———. 1974. *Legitimation of Belief*. Cambridge: Cambridge University Press.

———. 1979. A Social Contract in Search of an Idiom: The Demise of the Danegeld State. In *Spectacles and Predicaments*. Cambridge: Cambridge University Press.

———. 1985. Positivism Against Hegelianism. In *Relativism and the Social Sciences*. Cambridge: Cambridge University Press.

———. 1988. *Plough, Sword and Book: The Structure of Human History*. London: Collins Harvill.

Hall, John A. 2010. *Ernest Gellner: An Intellectual Biography*. London: Verso.

Polanyi, Karl. 1944. *The Great Transformation*. New York: Farrar and Rinehart.

Scheidel, Walter. 2017. *The Great Leveler: Violence and the History of Inequality from the Stone Age to the Twenty-First Century*. Princeton: Princeton University Press.

Sunstein, Cass R. 2017. *#Republic: Divided Democracy in the Age of Social Media*. Princeton: Princeton University Press.

CHAPTER 8

How to Integrate Economic, Social, and Political Theory: Revise the Rationality Principle

John Wettersten

Introduction

Economic theories today are all too often isolated from social and political ones. This isolation should enable economists to develop purely economic laws, which may then be applied in specific contexts involving other, non-economic factors. But the problem of how results of social scientific research which have first been separated by fields can then be combined in solutions to applied problems has no acceptable solution. Either results obtained through separation are not in any clear way applicable to real events, or realistic descriptions of specific areas of inquiry do not yield interesting theoretical results. It is, therefore, interesting to try to overcome the isolation.

This essay proposes a way of doing so. It does not criticize in detail standard approaches which use such separation, nor does it look at other attempts to reform methodology in economics in order to overcome what many have perceived as intellectual exercises with no clear connection to the real world—what Ronald H. Coase called "blackboard economics"

J. Wettersten (✉)
School of Social Sciences, University of Mannheim, Mannheim, Germany

R. Sassower, N. Laor (eds.), *The Impact of Critical Rationalism*,
https://doi.org/10.1007/978-3-319-90826-7_8

(1988) or what Hans Albert called "model Platonism" (1998, 114–126). Jeremy Shearmur (1996) and Lawrence Boland (2003) have made proposals for improving economic theory with the application of Popperian methodologies, even though they retain too much of standard equilibrium theory (Wettersten 2015).

Social scientific disciplines have a common ground in the theory of rationality. Since it is disunity rather than unity which is to be explained and overcome, this common ground may seem a curious place to look for understanding. But the pressures toward isolation found in various fields have the same source in the presumption that coherence and/or justification are minimal conditions for rational thought and action. By revising standards for rational thought and action, the rationale for the separation of economics from other social sciences may be overcome. No easy path to integrated solutions to social problems is thereby attained. But a research program for the effective partial integration of economics with other social scientific disciplines emerges. Under alternative assumptions about conditions which rational thought and action must meet, the *partial rationality of individuals* may not only be understood; it may also be used to both explain social events and provide proposals for social reform. The specific proposal for the revision of the rationality principle is that the rationality of individuals be explained as the *attempt to solve problems within given institutional frameworks* largely using established rules of thought has been developed elsewhere (Wettersten 2006). According to this proposal, institutions set the kind of problems individuals seek to solve and the desiderata they use to appraise solutions. Instead of explaining how institutions are produced by the rational actions of individuals, institutions are taken as given here, and their impact on how individuals pose and solve problems is studied. I intend to apply this view to both explain and overcome the separation of economic theories from social and political ones.

The Separation of Economic from Social and Political Theory

The separation of economic theories from social and political theories has one central theoretical cause: the desire to make economics, sociology, and political theory scientific. After social scientific tasks are completed in isolated fields, those interested in social planning are faced with the task of coherently combining the results. But attempts to combine results of various social sciences lead to ad hoc political decisions. And these may appear corrupt, because they are not based on universally applied standards. Special interests seem to take the place of agreed-upon standards. This unhappy result is in part an unintended consequence of attempts by economic, social, and political theorists to meet too high standards within their disciplines.

In their quest to make economics as scientific as possible, economists have sought to meet standards not matched by other social sciences. In order to do that, they isolate their own discipline in order to protect it from the dubious results of other fields. This strategy is determined by their aim of finding economic laws that are presumably describing causal relations between differing economic factors. A decrease in price will, for example, lead to an increase in demand, all other things being equal. The ideal of a complete system describing mathematical relations between all economic variables was formulated by Leon Walrás (Walrus 1965), though even he did not think it was possible to use this approach to describe the real world. In any real context, there are simply too many economic variables to include in one system. So, when building models, economists presume ideal coherent market systems in which relations between selected economic factors may be mathematically described. This ideal can be used to describe aspects of real systems. But just where and when they can be used to describe real systems depends on factors—both economic and noneconomic ones—not included in those models.

This program poses the problem: how many and/or which real causal factors can be left out of any ideal functional (coherent) system without destroying the ability to use it to describe real processes? The borders of economic science have been drawn and redrawn as new factors which influence economic processes have been added to deliberations. Coherent models of relations between increasing numbers of economic factors should lead to models which come closer to describing real-world situations. The movement to distinguish more sharply the science of economics was fuelled by opposition to thinkers such as John Commons (Commons 1934) and Thorstein Veblen (Veblen 1949) or the historical school in Germany. This movement was deemed necessary because those thinkers, who are now known as the old institutionalists, allegedly had no economic theory. As a result, they could offer historical and/or social analyses but no scientific laws. But as it became increasingly clear that new attempts to make economics scientific and capable of describing real social processes could not simply ignore institutions, new ways were sought to reincorporate them into economic models without rejecting the neoclassical approach.

The so-called "new institutionalists" sought to do better than the old ones by incorporating institutions into the new theoretical models. James Buchanan and Gordon Tullock (1965), for example, sought to explain the creation, maintenance, and change of institutions on the basis of the individualist presumptions of economic theory alone. I argued elsewhere that their efforts are empty: their results are compatible with the maintenance and/or formation of any institution at all (Wettersten 2006). They seek to provide methodological individualist explanations of institutions but cannot even do that because they cannot reduce institutions to a collection of actions based on rational decision-making. On their view, institutions pose cost-utility problems for the achievement of personal goals. Institutions will be formed when the cost of forming them is less than the utilities gained for those individuals who form them, and/or existing institutions will be used to achieve personal goals if the value of the individual's goals obtainable through joint action outweighs the costs. This approach enables economists to view institutional factors as products of the same calculations as (other) economic ones; their usefulness is merely part of the cost-utility computation of each (rational) actor.

But any model which includes such factors is still partial. This theory of the creation and use of institutions also does not help much in developing explanations of specific institutions, because it does not say how individuals will estimate the costs and utilities of joint actions. These costs are obviously quite different from prices, which can be clearly described. Institutions create problems or enable individuals to solve problems which are important to them but which are not all reducible to economic ones. The theory cannot account for these considerations other than to say that they are part of cost-utility computations of individuals. These are subjective and not accessible to empirical studies. Influenced by Joseph Schumpeter, Douglass North tried to fill a gap between economic theory and the history of economic activity (North 1989, 2005; Wettersten 2006). He began his career by applying ideas of Tullock and Buchanan to account for that history (North 1961). But by the end of the 1980s, he had, quite independently of Albert, come to the conclusion that Albert had reached in the early 1970s: a new explanatory approach was needed, if real economic phenomena were to be explained. And, like Albert, he has called for a new kind of explanation in economics which takes its cue from psychology (Albert 1972b; Wettersten 2006). It is unclear how this project is supposed to look, since psychology is far from offering psychological theories which can explain economic behavior.

Ronald Coase has tried to make models more realistic by adding transaction costs (1988, 1994). Firms and/or individuals do not merely react to prices, since different transactions may entail various costs. A higher price may be better if a lower price calls for more protection against deception, for example. Models which take such factors into account should be better able to make reliable predictions. But, even if additional factors are built into models, the models remain partial. The fundamental problem is that the influence of factors not part of any model remains unknown. This becomes clear in the attempt by Oliver Williamson to apply the idea of transaction costs to the problem of how firms are organized (1986, 1990, 1994, 1996; Wettersten 2006, 96–101). Williamson seeks to explain why firms choose a vertical organization in which they produce the goods and services they need for the production of some end product or why they choose a horizontal organization in which they purchase such goods and services from other firms. He argues that the choice depends on transaction costs. But such organization is surely also dependent on cultural settings. Stanford Jacoby (2005), for example, has shown how cultural traditions lead to different employment policies in America and Japan. Amartya Sen (2000, 2002) has pointed out the importance which individuals place on freedom and process (Wettersten 2009).

Some economists have also dealt more intensively with the obvious fact that politics influences how economic systems function. But they have had trouble reconciling democratic politics with free-market economies although this reconciliation should be easy, since both are individualist. Joseph Schumpeter (1994) suggested that capitalist societies are simply too risky for most individuals and as a consequence cannot survive democracy. Friedrich Hayek (1976) sought to isolate economics from politics even to the point of apologizing for Pinochet's dictatorship in the Santiago newspaper *El Mercurio*, because Pinochet secured economic freedom. Hayek also attempted to devise a political theory to protect economic policy from any too direct democracy. This attempt is quixotic, because it exemplifies a social planning in political theory that is rejected in economic theory, and anachronistic, because it flies in the face of modern movements to make democracy more direct.

Perhaps the most influential attempt to unify political and economic theory has been theories which explain the behavior of political parties and voters in democratic systems with rational choice theories (Downs 1957).

On this view, political activity is simply an extension of the application of economic methods to political affairs. The same kind of models used to explain and/or predict economic affairs can be applied to political affairs. There can be, then, no conflict. This attempt has grave defects. First, it cannot offer any interesting explanations of real political processes, such as why one candidate is successful and not another; it can only interpret. It has been noted that it cannot explain why people vote at all. It does not offer serious tests of the hypotheses given, as has been effectively argued by Donald Green and Ian Shapiro (1994). If we want enlightenment about political processes, we will have to turn elsewhere.

An alternative to theories which seek to incorporate social factors into economic models has been offered by Milton Friedman: economic models predict when they are applied in specific situations. But the judgment of just when and where these models may be applied so as to yield true predictions must be left to the inarticulate knowledge of trained economists (1953, 25). Although this theory is by no means enlightening about the impact of social factors on economic processes nor plausible in its claim that economists know intuitively how and when these factors are relevant without being able to explain why, it is the best stop-gap measure economists have been able to come up with.

The Rationality Principle as the Source of Difficulties

Any attempt to overcome the separation of economic theory from theories of noneconomic social factors may seem to be damned from the start. It is now conceded by everyone that wholesale economic planning is a fool's errand which regularly does more damage than good. Yet the maintenance of democratic societies requires that methods be available for resolving conflicts between economic and noneconomic interests. The conundrum caused by the apparent economic necessity to avoid social planning so that free markets may provide for economic well-being and the social need to engage in it to overcome environmental and social problems leads to contradictory recommendations. Economists by and large advocate keeping hands off social developments, while groups concerned with the environment, social justice, and other political problems demand just the opposite.

Although the dominant theory of rationality is shared by the relevant disciplines, it has led to sharp disagreement rather than to agreement in regard to social planning. Economists believe social planning to be damaging to the well-functioning of economic systems, and do so because of the theory that free-market systems satisfy demands of rationality. Social systems emerge which follow social scientific laws, because they are products of rational actions of individuals, that is, of the formation of coherent plans about how to achieve goals justified by the facts.

What is needed is a revision of the principle. When this principle is revised in a way that takes the fallibilism of rationality into account, difficulties can be lessened. I will, then, discuss why the rationality principle is the source of the separation, explain my proposed revision of it, and defend my proposal against the charges that it is trivial and/or useless.

Presumably only if thought is coherent can it be understood, and only if it is justified can it lead to predictable results. But, in order to predict, economists do not merely assume that coherent plans are justified but also that they are limited to the pursuit of economic goals. Given standard theories of rationality, it would not be possible to portray economic activity as predictable without this limitation. If one were to try to explain how the moral, social, or religious beliefs lead to actions in the pursuit of individual ends given the logic of the situation, no system can be described as emerging out of rational actions. This is so because the action of each individual might be partially explained as a consequence of purely economic considerations, but no generalization about how all individuals behave is possible; such generalizations are needed if the functioning of markets is to be explained.

With his fallibilist theory of science, Popper initiated the development of the theory that rational thought is not justified thought. But he maintained high standards of coherence for rational thought which he deemed necessary for explanations in the social sciences. His own version of the rationality principle is the standard view minus justification. He knowingly attributes a higher level of rationality to individuals than they possess, because he thinks that only then is explanation in the social sciences possible. Ian C. Jarvie and Joseph Agassi improved Popper's theory of rationality (1957, 1962, 1985) by noting that rationality is normally partial, but nevertheless progressive. If we adopt their proposal and take rationality to be a matter of degree, we can explain action as rational without assuming the full coherence and/or the justification of the actor's beliefs. The rationales for functionalism, which assumes the coherence of social systems, and methodological individualism, which presumes the coherence of individual belief, collapse. And with them both extreme attitudes toward social planning collapse. These extremes are the theory that social planning should be used to control all elements of the society in order to create a coherent social system and the theory that social planning is to be completely avoided so that the rational action of individuals alone (pursuing their aims with coherent and justified plans) will lead to a well-functioning market system. We can neither plan with assurance that we are on the right path nor can we leave things alone and expect things to work out right because individual actions are rational.

According to the rationality principle, all rational actions are consequences of what goes on in the heads of individuals. From one perspective, this is trivially true, since without specific thoughts there can be neither institutions nor rational action. But it is nevertheless false, insofar as it presumes that the contents of thought are above all results of individual deliberations. The far larger part of the contents of thought consists of mere expressions of social situations which are adopted by individuals willy-nilly, as Mary Douglas (1987) has so forcefully argued. If we revise the rationality principle to take this fact into account, we can overcome the dichotomy between the strict rejection of planning in economics and the strict endorsement of planning in politics. We do not need a large revision of the rationality principle, because a small revision has wide consequences.

Instead of maintaining that all actions are the result of the beliefs of individuals, we may view all actions as the result of attempts by individuals to solve problems within institutional contexts. Under this revised assumption, we may pose questions which are not simply economic, simply social, or simply political. Asking which problems individuals seek to solve in specific institutional contexts and how they may be aided or hindered by institutions in their pursuits to solve them, the problems which politicians and economists try to solve turn out to be of the same kind. They are problems of how institutions guide the choice of problems and the intellectual methods used to solve them. Problems of what social consequences ensue when individuals pose problems need to be posed as well. We may also overcome the dichotomy between cognitive and social psychology: all cognitive activity is social, and much of social activity is cognitive (Wettersten 2014).

We start like anthropologists by looking at institutionalized views and how they shape the way individuals view their problems and potential solutions. We cannot follow functionalists who seek to describe the whole intellectual framework of some society and how it determines the lives of its citizens. Rather, we look at this or that institution, at businesses, schools, unions, churches, or combinations of them to see how they typically shape problems to individuals. We do not treat institutions as actors. Agassi, following Popper, has emphasized: societies do not have aims, only individuals do (1960, 1987). But we may see how institutions shape aims, problems, and desiderata for solutions to problems. This approach avoids assuming the coherence or justification of individuals' views, but as rational to a degree. Individuals may try to solve some problems, while meeting incompatible socially established standards imposed by their religion or profession. But we may still explain their moving from one standard (or context) to the other as rational to a degree by explaining how they avoid conflicts by switching from one belief system to the other as occasions demand. We may study how individuals integrate or fail to integrate various aspects of their lives within institutional constraints.

We may continue with questions about how institutions influence the course of events, be it economic, social, political, or some combination of the three. Institutions lead individuals to make assumptions, because individuals identify with them. Chosen actions are combinations of various institutional influences, but they do not serve to integrate the society. They only allow individuals to take the various aspects of their immediate institutional problem situations into account. The fact that Jews in Christian societies or Chinese in Muslim societies have to take account of the religious views of their fellow citizens does not mean that the result is some functioning whole; but it also does not mean that we cannot see their action as rational, as they use the institutional means at their disposal to solve their problems.

Social scientists may choose problems in broad or narrow ways, depending on their interests. Ernest Gellner (1974), for example, proposed to study how beliefs are legitimized in Western culture. But Gellner does not go far enough. Having started with such an anthropological study of how standards are used, one may proceed to study how individuals make use of such standards, what problems they face when they do, and what social consequences result from their attempts. We may also discuss how such standards may be reformed to improve the search for truth (Wettersten 1979). In this way, we are not simply studying the consequences of individuals' beliefs and actions but rather the consequences of the use of established doctrines by partially rational individuals. They are partially rational, because they will rarely have comprehensive and often not even coherent views. Yet they do act on understandable presumptions, given their institutional contexts; their actions can be understood as attempts to solve problems in these contexts; and social events can be explained as consequences of them. One such study has offered enlightenment about extended families in modern societies by examining which tasks they carry out (Lettwark et al. 2003; Wettersten 2007). Social scientific explanations are always piecemeal to this or that degree: they explain how individuals use institutional contexts peculiar to their social positions to pose problems and to solve them. Their problem-posing and problem-solving activities influence other individuals who may use different assumptions for whatever reasons. Such influences and interactions may be happy ones when, for example, individuals from one culture achieve economic success in a new context but can also be disastrous when, for example, individuals are excluded from participation in society because of their religious activities.

Why It Is Better to Study Institutions Rather than Situations

The proposal made here changes the task of the social sciences from that of the study of the logic of various situations—the approach followed in individualist theories of social scientific method today—to the study of institutions. Some may find this to be an unfortunate switch, because it allegedly narrows the scope of social scientific research without enriching it. This critical comment ignores two central advantages which the proposal possesses. First, it solves how one can identify the "logic of the situation" in any specific case, a central problem which defenders of the established view have not been able to solve in any adequate way (because it cannot generalize to all cases). Second, its application may also be broader than that of the traditional view, because it does not restrict itself to those cases where a rational actor has some clear-cut and coherent response to a situation. Conflicting and irreconcilable demands may be studied. Functionalism may be rooted out of social scientific method without destroying the ability to understand social processes.

One reason that the problem of identifying a "logic of a situation" arises is the fact that this alleged logic is always relative to the (subjective) perception of a situation by the person acting. It is a trivial fact that different actors view the same situation in quite different ways. It is only by adding quite unrealistic assumptions about the aims and rationality of individuals that one can get anything like a clear specification of the logic of a situation; one may also run the danger of relapsing into unrealistic functionalist explanations. In order to avoid this fate, one can specify only some situations which are sufficiently alike to count as repetitions of the same logic. But defenders of this view have never been able to specify how similar and in what ways two events have to be similar to count as repetitions of the same logic. This has led to the wide use of Friedman's thoroughly inadequate suggestion that trained economists know intuitively which situations are similar enough to count as the same.

In contrast to this approach, we may start with institutions and study the rules they impose on individuals as they seek to solve problems. Such institutions and rules are relatively easily identifiable. We may form testable theories about institutions to see whether some formulations of the rules they impose vary or not, and, if they do, how. When these rules are laws formulated by governments, they may be presumed to apply in specified ways, unless we find corruption or lack of adherence. Another advantage of the use of this proposal over the use of the established rationality principle is that it is not limited to situations which are specified as posing some problem with a specific rational response, as is expected with the established rationality principle. This may be used to construct tests to see whether these responses are really made in specified contexts. But there is no path to finding out why they do not, if they do not. For this task, other methods are needed. My proposal may study how institutions lead various individuals to different paths and/or how they block various actions. And we may do so under the assumption that individuals are rational, not in the sense that they construct coherent plans for the pursuit of personal aims but as they pose and seek to solve problems in institutional contexts in accordance with institutional rules.

This proposal, then, allows social scientists to avoid falling back either into a Weberian functionalism which finds the whole of some society to be a product of a particular thought system or into the piecemeal functionalism of economic theory which sees various logics of situations as constituting closed systems maintained by rational actors pursuing their own financial gain in specific situations. We may study how institutions are dysfunctional, even when or especially when individuals are partially rational with their struggle to solve problems.

Conclusion

Problems in economic, social, and political theory may be solved using the same methods of explaining how institutions lead individuals to pose and solve problems in particular ways. This is possible because economic, social, and political institutions set rules which individuals by and large must use to pose problems and to choose desiderata which acceptable solutions must meet. Various social scientific endeavors may thereby be integrated. This integration resolves problems in the construction of social scientific models which arise due to the arbitrary selection of factors built into them. It may also overcome the persistence of functionalist assumptions in the social sciences long after such assumptions have been shown to be deleterious to the understanding of real social processes.

This essay is a revised version of a lecture given at the Walter Eucken Institut, Freiburg Universität, Germany.

References

Agassi, Joseph. 1960. Methodological Individualism. *British Journal of Sociology* 11: 214–270.

———. 1975a. Institutional Individualism. *British Journal of Sociology* 26: 144–155.

———. 1975b. Methodological Individualism and Institutional Individualism. In *Rationality: The Critical View*, ed. Joseph Agassi and Ian Jarvie, 119–150. Dordrecht: Martinus Nijhoff.

———. 1977. *Towards a Rational Philosophical Anthropology*. Dordrecht: Martinus Nijhoff.

———. 1988. Neo-Classical Economics as 18th Century Theory of Man. *Fundamenta Scientiae* 9 (2/3): 189–202.

———. 1991. Bye-Bye, Weber. *Philosophy of the Social Sciences* 21 (1): 102–109.

———. 1998. Knowledge Personal or Social. *Philosophy of the Social Sciences* 28 (4): 522–551.

———. 1999. *Liberal Nationalism for Israel*. Jerusalem: Gefen.

Albert, Hans. 1972a. *Ökonomische Ideologie und politische Theorie*. Göttingen: Verlag Otto Schwartz & Co.

———. 1972b. *Konstruktion und Kritik*. Hamburg: Hoffmann und Campe Verlag.

———. 1977. *Kritische Vernunft und menschlichen Praxis*. Stuttgart: Philip Reclam.

———. 1978. *Traktat über Rationale Praxis*. Tübingen: Mohr Siebeck Verlag.

———. 1986. *Freiheit und Ordnung*. Tübingen: Mohr Siebeck Verlag.

———. 1998. *Marktsoziologie und Entscheidungslogik*. Tübingen: Mohr Siebeck Verlag.

Boland, Lawrence. 2003. *The Foundation of Economic Methodology: A Popperian Perspective*. London: Routledge.

Buchanan, James M., and Gordon Tullock. 1965. *The Calculus of Consent*. Ann Arbor: University of Michigan Press.

Coase, Ronald H. 1988. *The Firm, the Market and the Law*. Chicago and London: University of Chicago Press.

———. 1994. *Essays on Economics and Economists*. Chicago: University of Chicago Press.

Commons, John R. 1934. *Institutional Economics*. New York: Macmillan.

Dahrendorf, Ralf. 1974. *Homo Sociologicus*. Opladen: Westdeutscher Verlag.

Douglas, Mary. 1987. *How Institutions Think*. London: Routledge & Kegan Paul.

Downs, Anthony. 1957. *An Economic Theory of Democracy*. New York: Harper & Row.

Friedman, Milton. 1953. *Essays in Positive Economics*. Chicago and London: University of Chicago Press.

Gellner, Ernest. 1974. *Legitimation of Belief*. Cambridge: Cambridge University Press.

———. 1996. Reply to Critics. In *The Social Philosophy of Ernest Gellner*, ed. John A. Hall and Ian C. Jarvie, 670–672. Amsterdam and Atlanta: Rodopi.

Green, Donald P., and Ian Shapiro. 1994. *Pathologies of Rational Choice Theory*. New Haven: Yale University Press.

Harrison, Lawrence E., and Samuel P. Huntington, eds. 2000. *Culture Matters: How Values Shape Human Progress*. New York: Basic Books.

von Hayek, Friedrich A. 1976. Economic Freedom and Representative Government. In *Royal Society of Arts*. London: Institute of Economic Affairs.

Jacoby, Stanford M. 2005. *The Embedded Corporation*. Princeton: Princeton University Press.

Lettwark, Eugene, Merrill Silverstein, Vern L. Bengston, and Ynez Wilson Hirst. 2003. Theories About Families, Organizations, and Social Supports. In *Global Aging and Challenges to Families*, ed. Vern L. Bengston and Ariela Lowenstein, 27–53. New York: Aldine de Gruyter.

North, Douglass C. 1961. *The Economic Growth of the United States, 1790 to 1860*. Englewood Cliffs, NJ: Prentice-Hall, Inc.

———. 1989. Comments. In *The Economic Role of the State*, ed. Joseph E. Stiglitz, 107–116. Oxford: Basil Blackwell.

———. 2005. *Understanding the Process of Economic Change*. Princeton: Princeton University Press.

Popper, Karl R. 1957. *The Poverty of Historicism*. London: Routledge and Kegan Paul.

———. 1962. *The Open Society and Its Enemies [1945]*. New York: Harper & Row.

———. 1985. The Rationality Principle. In *Popper Selections*, ed. David Miller, 357ff. Princeton: Princeton University Press.

Shearmur, Jeremy. 1996. *Hayek and After: Hayekian Liberalism as a Research Program*. London and New York: Routledge.

Schumpeter, Joseph A. 1994. *Capitalism, Socialism and Democracy*. London: Routledge.

Sen, Amartya. 2000. *Development as Freedom*. New York: Random House.

———. 2002. *Rationality and Freedom*. Cambridge: Harvard University Press.

Stiglitz, Joseph E. 2002. *The Economic Role of the State*. Oxford: Basil Blackwell.

Vanberg, Viktor. 2002. Rational Choice vs. Program-Based Behavior. *Rationality and Society* 14: 7–53.

———. 2004. The Rationality Postulate in Economics: Its Ambiguity, Its Deficiency and Its Evolutionary Alternative. *Journal of Economic Methodology* 11 (1): 1–29.

Veblen, Thorstein. 1949. *The Theory of the Leisure Class: An Economic Study of Institutions*. London: Allen & Unwin.

Walrus, Léon. 1965. *Elements of Pure Economics: Or the Theory of Social Wealth*. Translated by William Jaffé. London: Allen and Unwin.

Wettersten, John. 1979. Ernest Gellner: A Wittgensteinian Rationalist. *Philosophica* 8 (4): 741–769.

———. 2006. *How Do Institutions Steer Events? An Inquiry into the Limits and Possibilities of Rational Thought and Action*. Aldershot: Ashgate Publishing Co.

———. 2007. Philosophical Anthropology Can Help Social Scientists Learn from Empirical Tests. *Journal for the Theory of Social Behavior* 37 (3): 295–318.

———. 2009. Popper and Sen on Rationality and Economics: Two (Independent) Wrong Turns Can Be Remedied with the Same Program. In *Rethinking Popper*, ed. Zuzana Parusniková and Robert Cohen, 369–378. Netherland: Springer.

———. 2014. New Social Tasks for Cognitive Psychology; or, New Cognitive Tasks for Social Psychology. *The American Journal of Psychology* 127 (4): 403–418.

———. 2015. Critical Rationalists Much Too Narrow Contribution to *Der Positivismusstreit*: Alternative Political Theories Which Conform to Fallibilist, Individualist Methodologies Were Needed Then and Are Needed Today. *Journal for Classical Sociology* 15: 102–114.

Williamson, Oliver E. 1986. *Economic Organization: Firms, Markets, and Policy Control*. Brighton: Wheatsheaf Books.

———. 1990. *Industrial Organization*. Hants, Broomfield, and Vermont: Edward Elgar Publishing Ltd.

———. 1996. *The Mechanisms of Government*. New York and Oxford: Oxford University Press.

———. 1994. Transaction Cost Economics and Organization Theory. In *The Handbook of Economic Sociology*, ed. Neil J. Smelser and Richard Swedberg, 77–107. Princeton: Princeton University Press.

PART II

Critical Rationalism

CHAPTER 9

The Future of Critical Rationalism

Joseph Agassi

As some would argue, "novelty in philosophy is often a matter of daring rather than invention" (Wolfson 1934, p. 331). Writing an intellectual forecast is breaking a major taboo of the critical rationalist school of philosophy—avoid long-term forecasts, since there is no plan for historical processes: the future is open. This maxim holds particularly for intellectual processes, except that this school also respects no taboos. The only ideas or rather practices that distinguish this school comprise the willingness to invite, welcome, encourage, and promote presentations of problems, of possible solutions to them, and of criticism, regardless of their origins and the possible biases in their backgrounds. Ian C. Jarvie has overlooked the taboo when writing about the future in his majestic "Utopian Thinking and the Architect" (1986, pp. 328–351) in which the architect serves as an archetype for a planner of the distant future. How can we plan for the future, knowing that the future is open? How do architects plan houses that are supposed to stand and function centuries from now? The answer is, poorly: we always plan in ignorance. This does not absolve us from efforts to imagine the future and to plan our projects in manners that

J. Agassi (✉)
Department of Philosophy, Tel Aviv University, Tel Aviv, Israel

York University, Toronto, ON, Canada
e-mail: agass@post.tau.ac.il

R. Sassower, N. Laor (eds.), *The Impact of Critical Rationalism*,
https://doi.org/10.1007/978-3-319-90826-7_9

encourage improvements of all sorts. And one way to improve our guesses is to try to criticize our assumptions about them and to offer alternatives to them. Thus, architects may devise ever newer kinds of houses; they may plan houses that are easy to alter and even to replace. Indeed, all things being equal, increasing the ease with which a house is alterable seems desirable. This, of course, is not the case of societies with conservative attitudes to architecture.

Karl Popper's proof that the future is open rests on the idea that science has a great influence on all plans and that possible future discoveries are obviously unknown today. True, yet future practical problems are possibly soluble today and remain in the future. Even known solutions of current pressing problems usually await their implementations (e.g., green housing). Most daring social implementations of solutions lag. Good planners know that. An obvious example is when legislation comes to solve pressing problems, letting other problems wait for their turns. Orders of priorities are thus most important; they may be regrettably erroneous: urgent problems may seem not pressing and vice versa; similarly, solutions to urgent problems can be too short-termed or too long-termed. And this kind of assessment may be erroneous as well. Notoriously, legislative errors are not always easy to remedy: it often takes too much time. This serious problem afflicts, in principle, all legislation. It is therefore wise to make a constitutional law that as a default position all laws should be on trial for a fixed period ("sunset"), to be put automatically on the agenda for a reasonable discussion afterward, to be then reinstated or altered or (by default) abolished. This idea needs much thrashing out and is then submitted to some legislative body for trial. This happens occasionally but not often enough. This is incentive for writing science fiction: authors with utopian ideas or utopian thinkers with some writing skills may give them some thought. And science fiction is known to influence planning on occasion. Not in vain did Isaac Asimov consider this sort of literature social science thought experiment.

Diverse future activities—planning, legislation, construction—are open for public discussion, including those of critical rationalists. Self-declared critical rationalists will naturally welcome such discussions, provided they are directed at problems we judge important or interesting. In case we disagree about this, we can try to put our disagreement on our agenda. (We may fail, of course; there is no guarantee on such matters.) Preferably, items on our agenda will be accompanied by the presuppositions that we expect the discussion to rest on, without shielding these presuppositions from critical examination, no matter how obvious these appear. (Again, we may overlook these, even though unwisely: there are no guarantees for wisdom.) The candidate for the default problem is simple: of the greatest needs of the community or society in question (the current philosophical community, in our case), which ones are urgent so as to deserve efforts to alleviate now or in the very near future? Before attempting to answer this question, we may want to critically examine our presuppositions. Famously, failures of predictions often happen due to oversight of unnoticed presuppositions, ones that take place as parts of tests of scientific theories or as parts of deliberations that should lead to some plans meant to solve some pressing problems. Presuppositions behind expectations regarding critical rationalism may stand out when examining critical rationalism in action. This pertains both to the efforts of critical rationalists to go about their own business and to the communal, social, or political activities of individuals who happen to be critical rationalists. Thus, Ian C. Jarvie, a leading critical rationalist—to the extent that such an animal exists—who is the editor of a top philosophical periodical (*Philosophy of the Social Sciences*), may influence one way or another the atmosphere in the philosophical community or in the community of social scientists. What presuppositions enter assessments of the future of such activities?

The first presupposition is that humanity has a future to begin with. The future at issue is the near future, say, the next century or so. Thus, the presupposition is that humanity will last that long. Indeed, not only humanity but also human culture more or less as we know it, namely, what we call today western culture, and more specifically, the rationalist or pro-science culture of western-type, liberal-democratic nation-states. This presupposition and its related ones are already highly problematic, even under the assumption that they are true: as presuppositions, they are much too vague. Consider the supposition that human society will remain rationalist. What does this amount to? Is human society rationalist these days? Is Western society? How? We have a vast group of philosophers, sociologists, and anthropologists of culture and of science and other research groups and propagandists of all sorts whose self-appointed task is to observe rationality in action and to comment on it in real time, if not also to advocate rationalism and the rationalist way of life whatever exactly they suppose it is. Most of them are not sufficiently rationalist, not in the sense in which we critical rationalists consider ourselves and our deliberations rational: people, including philosophers, often cling to ideas that have already been effectively criticized and even without bothering to rebut this criticism or put the dispute on the public agenda for rational debate. Will the rationalist sectors of extant research groups grow or dwindle? More important, will the disposition of these rationalist sectors grow or dwindle (whichever is the case), and sway the public this way or that way? The wish to influence the agenda of the commonwealth of learning may demand of critical rationalists to examine each option and its possible influence on the future.

The question of the future of critical rationalism is a small part of the question: will the world in general be more rationalist in the future? Is society going to be more open or more closed? Are we are going toward a new Middle Ages or to more and bigger progressive revolutions? And we can ask: what are the most general preconditions for rationalism-progressivism? The answer we have today hinges on individualism, liberalism, and democracy. It is generally agreed that this answer is deeply linked with market economics: as Friedrich von Hayek said, freedom is indivisible, so that guarding freedom must include guarding economic freedom. Nevertheless, his claim that this amounts to unrestrained market is questionable; Popper said it is even inconsistent. Hayek (1960) responded saying, trusting the welfare state and the democratic checks and balances is too naïve. Of course, this dispute presents diverging views of the future of rationalism (Agassi 2016). This divergence also hinges on the dispute about the ability of intellectuals to influence politics.

This, too, is questionable, and for diverse reasons. The image of the society with a free market as Adam Smith had envisaged is ideal. He noted that the Hanseatic League somewhat resembles it (1776, p. 395). This is no proof that his plan is at all possible. As Ernest Gellner has noted (1996, Chap. 1), the image of individual preferences is too bland. As we are all ambivalent to some extent, it is simply impossible that even the ideal market mechanism will optimize the preferences of individuals, even if it were possible, which it is not, to satisfy everyone's interests as well as the national interest. In other words, (as Hayek has noted) the theory of consumers' preference is redundant, since the consumer economic theory discusses large-scale consumption patterns. This leaves us all possibly too ambivalent about proposals to render attitudes toward them rational. Moreover, even if it were possible for intellectuals to influence sociopolitical planning optimally, it is not clear that this will take place.

Popper agreed with the wise observation that philosophers find it harder to exert influence when they appeal to reason than to emotions, yet these are the cases of their significant influences: Popper agreed with the observation of Hegel and Heine that the French Revolution was born out of philosophical deliberations that had taken place a generation earlier. This seems to me too obvious for discussion: liberalism—the idea of Machiavelli that legislators should legislate with the intent of making it in the interest of citizens to be law-abiding—sounded too utopian when it appeared, but it is now all too obviously triumphant, seeing that the rich part of the world and the liberal part of the world coincide. It therefore never ceases to surprise me that so many intellectuals oppose it.

Liberalism was a major change the likes of which the world seldom saw (and the philosophical innovations of Rousseau and of Smith are inherently liberal). What intellectuals may still hope for is to affect much smaller but still significant changes. If some serious thinkers will study the ways intellectuals may cause small changes that may generate significant ones, then people who will try to implement the practical conclusions of these studies may influence the future in unimaginable manners. Indeed, some such studies already exist. The simplest example is the ability of philosophers to use existing institutional arrangements to implement social measures that might raise the ability of the public to be critical of the political activities of their leaders and to influence them. There is unexpected potential in the very combination of the idea of *participatory democracy* with the idea of Pericles that the most significant aspect of democracy is the ability of the citizen to criticize.

These considerations apply to the possibility of intellectuals to enhance democracy and freedom of speech, and to raise public interest rapidly at crucial moments. In particular, we may take for granted that the greatest impact of thinking people on social and political life in general is to draw attention to obvious and important matters that the public and the leaders are ignorant of or tend to ignore. An example is the British public response to the institution of purchase by installment (hire purchase). Public opinion then opposed it on moral grounds. The authorities instituted a campaign to inform the public that this measure was necessary to prevent massive unemployment. Other and more pertinent examples concern the legal limitation on the purchase of harmful commodities in the hope that this will reduce the harm their abuse causes. This was tried first in the USA with famous ill effects. Nor is it obvious that all constraints on purchase of dangerous commodities are welcome. Most of those who insistently demand the increase of measures to constrain the availability of drugs that are abused passionately oppose any effort to limit the availability of weapons (gun control). Perhaps a better understanding of the facts will help improve the situation.

There are much more problematic cases, and they are much more difficult to survey. Also, here personal bias may become too big to keep the disquisition valuable to any degree. They comprise discussions of the problems that beset critical rationalism and the way we may tackle them. Here the major argument against historical prediction (Popper 1957, Preface) comes into play: predicting future innovations makes them present already. Nevertheless, we have no taboos, and we can always indulge in science fiction. Some science fiction may influence real events in that planners, engineers, or social reformers may be familiar with ideas that science fiction promulgates; they may then find it sufficiently palatable to try to implement them. So here the bias of the discussion of the internal developments of the school of thought may be of positive value, if and to the extent that they may provoke responses, be these in assent or in dissent of some far-reaching proposals. The socially most valuable science fiction narratives, the social science thought experiments, describe people who have overcome difficulties that their authors find most aggravating. And then their fantasies may provoke some innovations. A well-known example is the fictitious narrative of a black president of the USA that was very helpful in the overcoming of racial prejudices sufficiently for trying out that idea. This, however, did not invoke any new ideas.

Fictitious descriptions of some possible futures may belong to the utopian literature that Popper denounced. He had strong arguments against them. We are fallible, and comprehensive plans for social improvement may harbor big errors. When engaged in large-scale reforms, we have to overlook obstacles, and these obstacles may comprise criticisms of our comprehensive plans that we thus tend to postpone discussing for a while, and in the meanwhile, the utopian planner causes real suffering in exchange for mere dreams. So, Popper suggested replacing utopian social engineering with piecemeal social engineering that should tackle the most urgent problems around. This is noble, yet it invites critical comments. The comment of Jarvie pertains here: utopian thinking need not lead to objectionable action; it may be useful for piecemeal engineers. Paul Feyerabend has stressed that all ideas may be approached in the wrong spirit: even persecution in the name of science, he observed, is not unheard of. Thus, utopian and piecemeal planning suggest a middle-ground utopia as a solution for an urgent problem that is not quite comprehensive. Such are the many science fiction stories about societies similar to ours but without racism or sexism or hunger.

The newly created global politics offer new arguments. Global politics concerns practical problems not given to local solutions, but hopefully given to global solutions: they concern the survival of humanity. They thus look like utopian engineering and in a manner that Popper could not consider when he wrote his monumental *The Open Society and Its Enemies* (1945). Possibly, the emergence of the ability of humanity to destroy itself and the threat that it might do so defy the distinction between piecemeal and comprehensive social engineering. Yet all this only enhances Popper's warnings: we can always make mistakes, and in the case of the preservation of humanity, one small mistake may bring about extinction. So, better take more precaution, and put every plan to much more critical scrutiny than ever before. More we cannot do. This suffices as a challenge; it invites the establishment of a new kind of institutions for the control of global political reforms. Critical rationalists may have a major role to play in ways that are problematic for uncritical rationalists to contemplate. It may invite efforts to institute the idea that criticism is beneficial and that disagreements and debates about them may make the difference between our survival and extinction.

A few more presuppositions are smuggled here. The second and most obvious one is that (at least for the next two or three generations) the critical rationalist school will remain as a school and that it will not lose its identity across the board. This seems possible only if some of its members will develop new ideas interesting enough to study and apply, theoretically and practically. For example, we should have better ideas about the demarcation between rational and irrational exchanges or at least between more and less profitable ones (in diverse senses of profitability).

A third one is about the fate of critical rationalism to change with the change of the (democratic) public education systems: with the increase of toleration and the power of reason in public (including political) life. If any aspect of public education will improve (by solving some urgent problems), then one might expect the status of critical rationalism to improve significantly. Even such a small change in the academic system might do that, such as the replacement of lecture courses with recorded lectures plus tutorials and seminars. As democracy moves between the Scylla and Charybdis of elitism and populism, the creation of democratic safeguards that should help politics steer away from both might already bring about a vast increase in the readiness to engage in critical debates on vital issues. Even the prevention of the many instituted conflicts of interests might do that, such as the one inherent in the accounting profession where accountants are paid by organizations to supervise them, in medicine where diagnosticians who recommend surgeries are allowed to perform them, and in the academy where professors grade exams of their own students. One of the worst conflicts in philosophy comprises the structure of appointment committees that decide the fate of philosophers of science: these include some science professors who are seldom good scientists, and so they tend to defend the public relations of science, and so they recommend appointing philosophers whose appointments look good to them. Consequently, leading philosophers of science in top universities are engaged in sheer public relations. It is hard to see how this situation may improve. Perhaps the Nobel committee may do something about it. Not that their activities are useful: they tried for many years to restrain the authority of the Chicago School of Economics with little success if any.

This is not to expect that philosophers will soon be much more advanced than they are now. Unlikely are reforms that will enhance the value of clarity of presentation of problems and problem orientation and criticism. Yet critical rationalists should help improve matters. Centuries ago Condorcet said that the future must be rosy since naturally every successful invention will be emulated. As this prediction was amply refuted, it may be useful to seek ways to limit the conditions that advanced the refuting instances. As Bertrand Russell has observed, in our time we have the choice between hell and a utopia our forefathers never dreamt of. He seems to have suggested that we have reached a stage in which we can hardly expect any situation in between the extremes of the very best and the very worst. Although an exaggeration, it is something to think about, not only because these extreme situations, hell and utopia, are only now available options, but also because they render the middle options very much less likely.

One of the greatest and most noticeable contributions of intellectuals to public life is their ability to devise new solutions to extant urgent social problems. These are often overlooked, especially since it is their implementers who take the credit (not without justice) rather than their originators (whose memory is often rewarded otherwise). In efforts to free a system, it is better to examine a new option critically than to accept it out of despair, but that is not always easy. For example, incredible as it sounds, one may devise an effective method to stop a running inflation before devising means to get the system out of the condition that has caused it. For a desperate move to be publicly accepted, the public must agree that the situation is indeed hopeless. We can go further: it may be possible to open a public discussion about a major obstacle that keeps a system stuck, even with no good idea about any means for overcoming the obstacle. Consider the philosophical community. Popper deemed commitment to inductivism the greatest obstacle to its ability to contribute to knowledge and to well-being. As he argued with them with ever newer arguments, he was in error: he was trying to prove to them that induction is no good, and this they took it for granted, and not unreasonably, that attainment of academic positions as philosophers of science requires of them to defend science, and the easiest way to do so is to defend some inductivist philosophy or another, as William Bartley suggested, a reasonable version of justificationism. The question is then dual: what makes it so difficult to avoid justificationism, and how is it possible to overcome this difficulty? Mario Bunge took it for granted that the study of science and the learning to appreciate science will cure one of inductivism, perhaps with the help of philosophical studies such as Bunge's own.

Ian C. Jarvie edits a periodical, *The Philosophy of the Social Sciences*, that encourages criticism and that plays down inductivism with no fanfare. There are many creative scientists who are highly critically minded, even philosophically minded, whose inductivism is nevertheless quite unshakeable. My paradigm case is leading physicist Max Born, who argued vehemently against Popper's critique of induction. Despite his inductivism Born was a friend of Albert Einstein who corresponded with him on scientific matters. This shows an unusual intellectual ability. In a sense, this holds for Russell, too, although his inductivism or sensationalism was but a leftover from a previous era. Nevertheless, despite the enormous popularity of Popper's political works, the first significant rise in the appreciation of critical rationalism happened among researchers in the natural sciences. This should encourage the development of tools for the social sciences, such as Jarvie's periodical (and collections of essays and similar activities). We may consequently see more individuals in public positions in the philosophical community with a better grasp of and interest in critical rationalism. Yet this should not be a proselytizing campaign. To the extent that the rise in the understanding of critical rationalism, with or without assent, improves research, this process is advantageous, as well as to the extent to which this process will bring about some valuable criticism of it. And, since criticism is unpredictable, so is the future of critical rationalism.

Let me be bold and refer to a current serious problem and to the possible critical rationalist solution to it. The difficulty in finding such an example is philosophical: critical rationalism is a minimal program, so that whatever it requires uncritical rationalism requires too, but not the other way around. Nevertheless, the lower expectations from the application of critical rationalism may be beneficial. Consider the case of high expectations, then. These days there are great disappointments from referenda. Both the referendum on a constitution for the European Union and the one of Britain leaving the Union (Brexit) led to disappointments. The authorities used the referenda with insufficient preparation and care. The recognition that rationality is limited will make them more cautious and more aware of the limitation of rationality and of democracy.

The most valuable lesson from critical rationalism is an idea that already Bernard Shaw advocated as sheer commonsense: the best way to approach any problem situation is by playing the devil's advocate, regularly and in every possible way. This, Shaw observed, reduces individual bias as much as is reasonable to expect: it is the best way we know for opening new horizons. The future is open, yet it is our duty to be optimistic, said arch-critical-rationalist Popper. As to assessments of what the future bodes for us, since life is a system of mostly lost opportunities, as any side-glance at political history show, it is hard to judge whether optimism or pessimism is nearer the mark. Yet as far as intellectual history is concerned, and so also as to its influence on future quality of life in general, there is ground for cautious optimism: if and when the dust settles, the picture often gets clarified to the benefit of all. (Popper struggled with physicists about some philosophical problems on the borderline with physics. Not being a physicist, he was at a disadvantage. He would have been delighted to learn that the criticisms that he voiced are now often, not always, taken for granted.)

There is here a mistaken system of acknowledgment. Since the scientific revolution, acknowledgment goes to discoverers of new facts and to initiators of corroborated ideas. This is unjust to critics other than discoverers of new facts, to conveners of all sorts, editors of periodicals, encyclopedias, organizers of conferences, and to authors of overviews. (This is due to the humility of Robert Boyle, who was the arch convener and who wrote the standards that are received to this very day with almost no change.) And Ian C. Jarvie excels not only as a philosopher but also, in his indefatigable work as an editor, as a convener and also as an author of significant overviews.

Acknowledgements My gratitude to Nimrod Bar-Am and Raphael Sassower for their help.

References

Agassi, Joseph. 2016. Toward a Fictionless Liberalism. *Philosophy of the Social Sciences* 46 (1): 77–91.

Gellner, Ernest. 1996. *The Psychoanalytic Movement: The Cunning of Unreason*. Evanston, IL: Northwestern University Press.

von Hayek, Friedrich. 1960. *The Constitution of Liberty*. Chicago: University of Chicago Press.

Jarvie, Ian C. 1986. *Thinking About Society: Theory and Practice*. Dordrecht, Holland: D. Reidel Publishing Company.

Popper, Karl R. 1945/1962. *The Open Society and Its Enemies*. Princeton, NJ: Princeton University Press.

———. 1957. *The Poverty of Historicism*. New York: Harper & Row.

Smith, Adam. 1776/1937. *An Inquiry into the Nature and Causes of the Wealth of Nations*. New York: The Modern Library.

Wolfson, Harry A. 1934. *The Philosophy of Spinoza*. Cleveland and New York: Meridian Books.

CHAPTER 10

The Dark Side of Technological Progress

Mario Bunge

The Bright Side

Progress in a social sector consists in the replacement of some items with better–truer, fairer, or more accessible–items. For instance, the replacement of oligarchies was politically progressive, and so was the substitution of enlightened for superstitious procedures in medicine, and of humane with cruel methods in penology. Obviously, every advancement will hurt those who benefit from the old regime, but not everyone will benefit equally from novelties. Nobody doubts that technology has progressed at a phenomenal pace since about 1800. Suffice to remember some of the milestones passed in the last century: automobile and air travel, birth control pill, automation, informatics, Internet, nuclear weapons, personal computer, pharmaceutical drugs, plastics, radio, television, and weapons of mass deception and destruction.

All such innovations have been produced by private firms or governments employing armies of skilled workers and managers using science, engineering, and huge capitals. And most of the modern technological advances have been translated into mass-produced merchandises, some of them affordable to increasing swathes of the people around the world. The popular press has spread the belief that industry will keep producing

M. Bunge (✉)
Department of Philosophy, McGill University, Montreal, QC, Canada

R. Sassower, N. Laor (eds.), *The Impact of Critical Rationalism*,
https://doi.org/10.1007/978-3-319-90826-7_10

increasingly sophisticated merchandise. But the same press avoids telling us what to do about the ills of technological progress, such as unemployment, large-scale military aggression, and pollution. Obviously, we need a finer analysis and a more daring political and ethical evaluation of technological progress (Bunge 2018). Let us proceed to sketching such an examination.

The Engines of the Three Industrial Revolutions

Let us begin by distinguishing three phases of the technical evolution in modern times: the first (1750–1850), second (1850–1950), and third (1950–present) industrial revolutions. The first, or cotton cloth revolution, resulted from the combination of steam power with the venture capital produced by the slave trade and colonization. The second, or internal combustion revolution, resulted from a combination of gasoline-driven machines with the financial capital generated by the first industrial revolution and the exploitation of colonies and semi colonies. And the third stage is characterized by a combination of electronics and pharmacology with financial capital. In short, we must distinguish three icons of modern merchandise: cotton cloth, car, and electronic device.

What generated those revolutions? The first industrial revolution resulted from the cotton mills designed by ingenious craftsmen, like James Watt, operated by cheap labor, and financed by the slave trade (Beckert 2014). Science contributed very little to this development. In contrast, the second industrial revolution resulted largely from an application of classical electrodynamics and biochemistry. And the intellectual component of the third or present phase is a combination of electronics with robotics.

If we are to believe Max Weber, the ethical component of the first industrial revolution would be the ascetic aspect of Calvinism, that most somber of theologies. The great Adam Smith would disagree: he stated that the industrial captains were driven by selfishness. Joseph Schumpeter stressed entrepreneurship or risk-taking. And some of the economists who analyzed the latest phase of the economy have held that what moves great businessmen is greed.

However, when evaluating the worth of any social innovation, let us remember that, since the dawn of civilization some five millennia ago, social stasis and innovations have occurred in social systems rather than in isolated brains(Polanyi 2001). In particular, technological advances have occurred as suggested here:

Technology & Capital ← Government
↓ ↓
Industry → Consumers

The schema above represents technologico-industrial changes in inegalitarian societies. In such societies, governments are elitist, and citizens have no other civic duty than saluting the flag and voting every so often for candidates vetted and financed by the powers that be (Wolin 2008).

The Dark Side

People may be classed into technophilic and technophobic. While technophiles emphasize and laud the bright side of technological progress, technophobes stress its dark side, in particular unemployment, alienation, pollution, and political intimidation. Nobody can dispute that machines displace manual workers, particularly since the goal of much industrial innovation is to introduce laborsaving devices. Technophiles argue that this is compensated by low prices. Something similar is true of other negative aspects of technological innovation. In particular, since we know the causes of pollution, we can avert it by clever measures to manage garbage and other kinds of waste. Still, the fact is that little is being done to minimize the adverse aspects of technological progress. Any political analyst can tell us the reasons for such neglect: costs, both economic and political. Indeed, the defenders of free enterprise as well as the populist politicians reject the regulations that would prevent businessmen from using nature as a garbage can.

The obscurantists have stressed the dark side in question, while the progressives have countered with a defense of such modern values as rationality and the scientific stance. Such a strategy wins among the people imbued with the love of knowledge and rational action. But working and fighting for a good life involves much more than winning disputes. In particular, the excluded do not benefit from our showing that Heidegger and the other technophobes were wrong and hypocritical since they were complicit with Nazism, the worst enemy of reason and goodness. In short, minimizing the bad consequences of technological progress takes more than refuting technophobia. It takes designing and putting into place mechanisms of social control. Let us take a quick look at past attempts of this kind.

Some Past Attempts to Bell the Cat

Technological progress has always benefited only a small portion of society, and as a consequence it has been resisted by those likely to be hurt by it. Let us recall just three episodes of the anti-technological resistance movement. The first is the banning of all secular thinking by the Christian theocracy headed by the Byzantine emperor Justinian (527–565), famous for his victories over the "barbarians" and his efforts to codify Roman law but infamous for his brutal cultural censorship. His most radical and lasting edicts were those that censored all the intellectual and artistic activities, especially the technological ones. Those restrictions were the ones condemned by the earliest knights of laissez-faire policies, the physiocrats in the eighteenth century.

Our second example is the banning of oceangoing voyages and the destruction of big merchant ships by the Ming Dynasty around 1430. The reason for this measure is still being debated, but at the present time the most likely explanations are that the ruling emperor was afraid that the shipping magnates were becoming too powerful or that the contact with Europeans would soil Chinese customs.

The third landmark of this saga is Luddism, the short-lived movement of English workers (1811–1816) who reacted to the unemployment caused by the inception of cotton mills by destroying machines. This movement, suppressed by force, was doomed to fail because it hurt not only the new capitalist class but also the millions of consumers of the new products. The moral of this case is that technological progress can be halted for a time but not forever.

The rise of worker's unions in the last decades of the nineteenth century was a partial solution to the problem of technological unemployment. The unions, agrarian as well as industrial, became the backbone of the Swedish progressive movement, but have been decimated and rendered powerless in the USA, where only 7% of the working-class population has been allowed to unionize.

A Modest Proposal

I suggest that, since technological advancement is ambivalent, its negative consequences can be avoided only if the root of such ambivalence is found and corrected. I further suggest that the root in question is this: unlike progress in all other fields, technological progress is *owned* by the bosses, shareowners, or managers of the firms where such progress occurs. Eliminate the included/excluded barriers, that is, democratize the economy, and the benefits of technological progress may be reaped by all of its creators, workers, managers, and policy-makers. In turn, such a democratization can be achieved if cooperatives replace private enterprises. Shorter: the key to the significant shrinking of the negative consequences of technological progress is the socialization or democratization of the bulk of the means of production, financing, and circulation of merchandises. In other works, to defang technological progress, let us replace oligarchy by integral democracy, aka democratic socialism.

Conclusion

Technological progress is the only ambivalent progress, that is, the one that has a dark as well as bright aspect. The root of this ambivalence is social inequality, the fact that most of a nation's wealth is in the hands of the few. Introduce into the economic system the same equality enjoyed by the members of the political and cultural democracies, and the dark features of technological progress are likely to shrink. One could represent my proposal in the following manner:

Benign technology ← Democratic government
↓ ↓ ↑
Benign industry → Educated and participant citizenry

Technologico-industrial changes under integral democracy. Note that technology has been reduced to *benign* technology, government has been transformed into *democratic* government, and the consumer collection into the body of politically *educated* and *participant* citizens. The first change is affected by banning the arms industry, and the second by broadening government from parliamentary to participatory (Bunge 2009).

References

Beckert, Sven. 2014. *The Empire of Cotton*. New York: Alfred A. Knopf.
Bunge, Mario. 2009. *Political Philosophy*. New Brunswick, NJ: Transaction.
———. 2018. *From a Scientific Point of View*. Newcastle: Cambridge Scholars.
Polanyi, Karl. 2001 [1944]. *The Great Transformation*. Boston: Beacon Press.
Wolin, Sheldon. 2008. *Democracy Incorporated*. Princeton, NJ: Princeton University Press.

CHAPTER 11

Fractured Knowledge: "Fake News"

Jagdish Hattiangadi

The Politics of Illiberalism

Secular or free thought, inconvenient bits of knowledge, and the rule of law are under attack again. But this time the attack is not inspired by the infamous and intolerant ideologies from World War II. It is not from the left, led by the international socialists, headquartered in the USSR, and not from the right, led by the national socialists. Before the war, independent secular thought was respected mainly in the liberal democracies of Europe and the Americas. After World War II, liberal democracies have flourished, and so has secular thought. After the war, the US took over leadership of the world order from Great Britain. It led by proposing and helping establish liberal international institutions in order to broker peace between nations. These institutions included the United Nations and its subsidiary organs, such as the Security Council and the International Court of Justice, as well as the Nuclear Non-Proliferation Treaty. The World Trade Organization has promoted world trade and so boosted world income. Eleanor Roosevelt led the effort to adopt the Universal Declaration of Human Rights at the UN General Assembly.

J. Hattiangadi (✉)
Department of Philosophy, York University, Toronto, ON, Canada
e-mail: jagdish@yorku.ca

R. Sassower, N. Laor (eds.), *The Impact of Critical Rationalism*,
https://doi.org/10.1007/978-3-319-90826-7_11

The expression "liberal democracy" is being used to signify broadly a democracy that protects the freedom of speech and religion and the rule of law. The word "liberal" is also used sometimes to more narrowly represent freedom from government regulation, which is better called "laissez-faire liberalism." Supporting regulatory policies for a more equitable distribution of goods is a narrower allegiance also called "liberalism," for which the term "equitable liberalism" seems appropriate.

The current threat to liberal democracy, despite the resilience of US institutions, is slightly different from the threat during World War II. The recent referendum to exit the European Union in Great Britain and later the election process of an unorthodox president in the US have each shown in their respective countries a new level of dissatisfaction with international free trade, and with institutions promoting diversity and freedom. There appears to be a great dissatisfaction in the White House after the election, for instance, with many established institutions of government, including the courts of justice, the free press, and many international agreements.

There is a characteristic new movement in the US that has taken over the White House. It is evidently critical of the mainstream "elite," which is identified as the liberal democratic establishment (whether laissez-faire or equitable). This elite includes large segments of the news and entertainment media, colleges and universities, their professors and intellectuals, the federal bureaucracy, and most courts of justice. This new movement is distinguished by the fact that its members reject the existence of common knowledge across a perceived political divide. They reject, more particularly, any published facts that feel uncomfortable to them, calling them "fake news." If there is no common knowledge, as they claim, then of course knowledge on all important issues will be fractured along political lines.

We need to ask who supports fractured knowledge in the US, and why. An analysis of presidential voting patterns could shed some light on this. The votes in favor of the current coalition represent about a quarter of the electorate, or about half of those who voted in a close race. Three blocks among these voters are, first, traditional supporters of the Republican Party; second, economic nationalists; and third, religious conservatives who have been supporters of the Republican Party since Richard Nixon courted them. It is the third block which is dabbling in epistemology. In more detail:

1. The successful candidate was endorsed by the Republican Party which has long endorsed liberal democracy as a matter of principle. Republicans are therefore opposed in principle to the denial of

common knowledge, and of cries of "fake news"; their support for the coalition is therefore temporary, practical, and reluctant.

2. The second block includes three subgroups: those who lost jobs in the globalized economy, whose needs have not yet been adequately addressed; those who resent their lost standing in society, which has been eroded by globalization (as lower groups are catching up to them); and third, the captains of those industries which depend on the exploitation of American land and resources and have lagged in growth under globalization (and hence have trouble attracting capital). It is apparent that no subgroup in this second block has a special interest in fracturing knowledge, because the best case for their dissatisfaction is based precisely on common knowledge and widely acknowledged facts.
3. The third group rejects secular science and philosophy, common knowledge, and established facts, because it has come to believe that their religious rights under the First Amendment are at risk, and a divisive strategy alone can save them.

The last American election was too close for any party to neglect a block of votes. Hence even the doctrine of fractured knowledge has received official status.

On Science and Religious Politics

Theoretical science, as a specialized part of secular thought, does not in general clash with religious thought. In many diverse cultures and civilizations, science, or philosophy, does not clash with religion at all or rarely does so. Philosophy, which asks general questions of importance to human beings, and speculates about the answers to be found, is to be found in all cultures and all religious groups. Two religions, Christianity and Islam, stand out for having ongoing difficulties with secular thought, and in each case it was not the religions themselves that brought about their intolerance of philosophy. Christianity at first was like any other religion, but then it became an ideological emblem of the Roman Empire when Constantine legalized it and then adopted it as the state religion. As such, it could no longer tolerate free thought which could question its emblematic theses.

In a communal religion, the normal kind found even among the earliest human societies, members can identify each other by their common customs, language, and a network of kinship bonds. Rituals are shared, and whether a member believes them or not is hardly relevant to membership if they are already part of it. What is relevant is how people behave in key situations as a condition for continued membership. People who act in strange ways are in different degrees left out of communal events, or in severe cases formally excommunicated. But their belief is not what matters.

When Christianity became the state religion of the Roman Empire, however, it had a great burden to carry as the identifiable abstract emblem of a far-flung empire that was at war over many centuries with the equally vast Persian Empire to the East. These empires exhausted one another with their warring, and had hollowed each other out.

Constantine's new state religion was a way to hold together an empire. But it could not use kinship, common rituals and customs, or a shared language, to bind together the empire. All that was left in common was an abstract faith in God, in the gospel of Jesus in the New Testament, and in the Church itself. When any abstract belief is made an emblem, it can be used and reused to reinterpret pagan customs along Christian lines. This makes conversion less onerous. Christianity thus spread with comparative ease within the borders of the old empire, and beyond. People could accept an abstract belief, enjoy the protection of Rome, and keep their own language and customs.

The main casualty of this abstract form of empire is philosophy. When a religion becomes an ideological emblem, it becomes important to get the religious doctrine down exactly right. Soon after Constantine made the religion an ideological emblem, the first Council of Nicaea was called, and Christian orthodoxy came to be debated, determined, and then enforced across the empire. Philosophy then became dangerous because it can question orthodoxy and offer heretical speculations. This endangered the unity of the ideological empire. We are not surprised when Edward Gibbon tells us that the Christian Emperor Theophilus destroyed the Great Library of Alexandria. On the other hand, if the good Emperor did not succeed entirely, then perhaps Muslims succeeded when they won back Alexandria for their own empire.

Islam arose in a similar manner to Christianity in Persian Asia. A powerful family that had opposed the prophet Muhammad during his lifetime (the Umayyad) became the new Khalifat and soon embarked on reconquering the old lands of the Persian Empire in the name of the new emblematic religion, Islam. Ideological Islam was able to contain the spread of Christianity in the East by getting Christians to pay an additional tax in the territories that their ideology controlled. Getting the orthodox ideology right became a preoccupation of the ideological Muslim religion, just as it was in Christianity. Secular thought was also a casualty in an Islamic Persian Empire as it had been in Christian Rome. The Church Militants and Jihadi Islamists were soon fighting again, as had the Romans and the Persians, but now on behalf of their true religions. In each religion there are stories of how early practitioners destroyed libraries and killed philosophers and suppressed secular academies. Thus the Emperor Justinian, who codified Roman jurisprudence, also banned the last iteration of Plato's Academy. Similarly, the great secular University of Jundishapur in Sassanid Persia was destroyed by the Islamic empire.

There are also many tales of how these religions carefully nurtured science and philosophy in their midst, tales often cited as counter evidence to their intellectual intolerance. This is not counter evidence. Each religion was opposed to secular thought or to free philosophy, which was suppressed. But each carefully nourished and encouraged submissive philosophy that could be used to bolster current orthodoxy. Thus we see the rise of two paradoxical new medieval philosophies: Christian philosophy and Islamic philosophy. Each is submissive to and carefully monitored by an orthodoxy. Emblematic ideology has been a great new invention for political statecraft and has survived the loss of empires. Abstract ideology can become nonreligious. It can still unify political organization, as in linguistic or ethnic nationalisms. It was only in liberal democracies, which took up a tolerance of religions and ideologies as their motto, that philosophy was free again.

On Political Philosophy and on Science

A liberal democratic form of government that was adopted in the US was inspired by the British model of limited monarchy in 1688, and by the associated political philosophy of John Locke. We must explore both the success and failure of John Locke's philosophy to gauge the current unhappiness with key liberal democratic institutions. Following the Reformation, conflict was rife between nations that took up some new form of reformed religion and those supporting the Roman Catholic Church. In England, however, the situation was different. From the time that Henry VIII had first broken off from the Church in Rome until 1688, there was a constant problem of succession, which varied from Anglican to Puritan to Catholic at different times. In 1688, after King James II proved to be very unpopular for overstepping the limits of his authority, Parliament invited William and Mary of Orange to the throne, but enshrined a Bill of Rights. Among other limitations, the couple had to accept a principle of toleration. This principle allowed individuals of the realm to exercise the freedom of their conscience to practice any religion. They adopted a form of Locke's principle that they should tolerate every faith, with exceptions. Locke included papists and atheists among those who were not to be tolerated, because the former would be a threat to British safety and the latter would not be trusted to fulfill a contract. In time these exceptions were removed in Britain.

Locke's philosophy of tolerance distinguishes between those things that are known to us on the basis of natural reason and those based on revelation. In this respect, it differs from an earlier principle of toleration promoted by James II to protect Catholic interests in England. Revealed knowledge depends on faith, which in turn is a matter of differing opinion. What is known to us on the basis of reason is enforceable as part of the civil law of the land, but matters of revelation or faith are items for us to tolerate in others, concerning which we can agree to disagree.

The First Amendment in the US enshrines the principle of toleration by formally separating Church from state: "Congress shall make no law respecting an establishment of religion, or prohibiting the free exercise thereof; or abridging the freedom of speech, or of the press; or the right of the people peaceably to assemble, and to petition the Government for a redress of grievances." All religions are tolerated under a neutral civil authority which is the legislature, and this freedom is enforced by the executive under the watchful eye of the judiciary. This allows for a flowering of many different forms of religious life without any emblematic religion or ideology that defines the state itself. It is important to note that toleration is of religious faith, not of unreasonableness. Other liberties, such as freedom of speech, are also proclaimed. If one pushes a person off a ledge, one breaches the liberty of the other, and if this liberty is to be upheld in a court of law, one cannot claim a lack of faith in the principle of gravity as a valid defense of one's actions. Tolerance does not extend to the denial of common knowledge, including science, e.g., in a court of law.

A theory of knowledge is therefore one of the key constituents of Locke's principle of toleration. Locke proposed such a theory of knowledge which provided the scope and limits of human knowledge; he made a clear case for tolerance of articles of faith that we cannot know while endorsing those things that we do. Common knowledge (based on natural reason) allows us to adjudicate neutrally between the different ideologies and religions that are tolerated in a civil society. Locke did not endorse any ideology or faith but studied human knowledge as a medical practitioner may: just like human anatomy or physiology (Locke 1690). He argues against innate knowledge and suggests that all our knowledge arises from sense experience. When we confront the world, we learn about it in the content of our sensory images, the "simple ideas of sensation." Our language includes some very abstract words, but they, too, stand for ideas in the mind that are compounded from simple ideas of sensation. Everything more complexly stated is compounded from these sensory ideas. This gives us a criterion of knowledge: whatever is true in thought and language, and extracted by natural reason from sense impressions, is knowledge; anything which cannot be extracted by reason from the content of sense experience is at best an opinion. Theories about the true nature of God, or about life after death, for instance, cannot be known by reason from sense experience. They must count as matters of faith or belief which are to be tolerated.

An unexpected consequence of this account, which was unknown to Locke himself, was pointed out by David Hume (1777). All our sense impressions are gained locally and in particular times, whereas what we claim to know about the world can go beyond those times and places. If one claims to know that the sun will rise tomorrow, then one should be able to produce a chain of reasoning from my knowledge of past sense impressions of the rising sun to this statement about the future. But no such chain of reasoning is valid, unless we assume that the future will be like the past in this respect. If we call the method of reasoning from facts to generalities "inductive inferences," then no inductive inference can be valid without making illicit assumptions. Any first principle of a science like physics is therefore always a hypothesis, as Henri Poincaré put the matter early in the twentieth century (Poincaré 1905). *We always have a choice* concerning which hypothesis we adopt as a principle of physics. If we make a different choice, we can make appropriate adjustments in the rest of the system of theory to explain all the facts that we can now explain. When physicists claim that a certain principle of physics is true, this "truth" must be regarded as a *convention* among them: they could have adopted another equally adequate principle and called it "true" instead.

Formalist mathematicians writing about the foundations of mathematics at about the same time made another observation. Reflecting upon the discovery of the axioms of non-Euclidean geometries, they noted that the meanings of expressions like "straight line" are not the same in different axiomatic geometries. What intuitive sense can we make of the fact that there are many straight lines between any two points in some new geometry? All this makes much better sense if we imagine each system of geometry to be like a game, each of which has its own rules and axioms. Words used in these games ("straight lines") will mean what they do according to how their words behave, or are used. Formalism of this kind has an odd consequence: if we compare two systems of physical theory, then the facts that are described in one system will be untranslatable into the other. It seems this is so because what the facts of one system mean when stated in their own lexicon is internal to that lexicon or language, and it cannot be captured in another. Paul Feyerabend (1975) and Thomas Kuhn (1962) called this possibility "the incommensurability of theories." Since (on this account of meaning) there can be no lexicon common to two systems of theories (say Newton's and Einstein's theories of gravity), there can be no common facts that can be mutually or neutrally stated. We have no reason to say that one "explains the facts" better. Each system must be taken on faith, as it were. There are no common facts.

"No common facts" is a lot to swallow. Recently, a well-known film maker, Errol Morris, with a background in the history and philosophy of science, has apparently noted this similarity to our current predicament. John Horgan (2017) reports the claim that the incommensurability thesis has influenced the election of a "post-truth" presidency.

There are two difficulties with the "incommensurability thesis" of Feyerabend and Kuhn. First, while meanings of technical words are determined in part by their use in theoretical systems, it is a stretch to suggest that they are wholly so determined. The words "straight line," for instance, are not exclusive to geometry but also found in carpentry and mapmaking and much else. This alone undermines the claim that knowledge is fractured as claimed. Second, the critique of induction by Hume is also mistaken, which is harder to show; the critique assumes the truth of Locke's theory of knowledge, namely, that all our knowledge must arise by the use of reason from sense impressions. But Locke's theory is not the only one we could consider, though, if correct, it is well suited to his conception of government. The new theory of induction used by the Royal Society when Locke was a student at Oxford University, known as *Experimental Philosophy*, was based upon Francis Bacon's theory of knowledge (1620). It is this conception of knowledge (not Locke's, from 1690) that scientists like Robert Boyle, Thomas Sydenham, and Isaac Newton were putting into practice with renowned success.

Bacon thought of sense perception as a portal to the world around us, meaning that we perceive things through our senses. Perception does not act as a sensory veil between our understanding and reality, as Locke's later account inadvertently did. Bacon was convinced, however, that our minds, perceptions, languages, and theories distort the reality they transmit to us, and they do so in countless ways. His theory of induction was designed to show how we can nevertheless discover features of reality itself by removing some distortions. Had the mind but one distortion, its error would be undiscoverable, for we would interpret the world in its light. When there are many, they can conflict with one another, and come to our attention. If reality be distorted in two independent ways, where the errors conflict, we can use each distortion to challenge the other. But the sheer number and variety of errors of the mind are countless.

First, you take a small topic to study, suggests Bacon, and then create a "natural history" for it, that is, a written journal of errors apparent about its nature. Whenever you find the apparently true nature of a thing, you note down the conditions under which the thing so appears. Someone else should be able to follow your words to find the same nature. Then you find (this is "an experiment") the thing in a slightly different condition, showing a different underlying nature or reality. Again, you note the conditions of observation. A thing must be really what it is and not some other thing. So you have recorded a discrepancy in your history of its nature, or what Karl Popper later called a "problem" for our understanding (1934). You proceed to note down more and more such problems or difficulties, without trying to solve them immediately.

A whole nest of such problems will be recorded. They will seem impossible to solve in one go. What solves one problem will not normally solve another. Together they form what Kuhn later called a "puzzle," which is to be solved, all at once, with ingenuity (1962). Bacon liked to think of it as a "labyrinth" or sometimes as a cipher to be decoded or "interpreted." When a natural history is sufficiently paradoxical, there is a unique solution to its puzzles. A Baconian puzzle is always of this form: what must reality be like that it appears distorted exactly as it appears under the noted conditions? The model that you can work out to solve a puzzle is a thing of beauty when found. At this point your job is done, for the moment. This error analysis, when produced, constitutes knowledge of the underlying reality precisely because it constitutes a unique solution. It is a discovery, because it uncovers some noted distortions of our mind and reveals reality more faithfully than before. There are endless more distortions which may beset our thought, however, so this is just a first step and not the final truth.

Interestingly, we now have a criterion for judging reality—for judging that we have grasped reality better than before. This criterion is that we have a new "power over nature." If we recreate an experimental condition noted in our natural history (call it a "causal condition"), then we know exactly what distortion of reality (call it an "effect") we will find in the thing. Our unique model of reality tells us how to produce effects by producing the causal conditions we have noted down. Our *power over this bit of nature* gives us confidence that we have knowledge, that we have "got hold of the thing itself," that we are not merely producing empty words. Other errors may still lurk in our all-too-human understanding, but we do know something now about this reality itself. This method had resounding successes: Boyle's model of the sea of air, Sydenham's model of the epidemiology of disease, Newton's gravitational model of the solar system, and later, Antoine Lavoisier's model of combustion are the most notable early successes. Astonishingly enough, because Locke's conception of knowledge has the fatal flaw that we cannot rationally arrive at general knowledge, we have learned to interpret all Baconian knowledge as "hypothetical" or merely as opinion. It is not.

Baconian knowledge gives us a model with a little power over nature, but its extension always remains speculative. Apart from a limited model of a bit of reality, everything else remains a matter of opinion. Each principle of science is still a hypothesis, and each hypothesis is still full of error. Science remains, for the most part, tentative in its theory. It is for this reason that historians of science who want to find hypotheses in science have no difficulty in finding them. Opinions abound in science. Uniquely generated knowledge is scarce and found only occasionally among the many opinions.

If opinions were all that we had, then some questions would have arisen. Why is it that, of the many opinions, some are considered true and some only remote possibilities? Why is the received opinion received? This is of importance in politics. One influential recent way of answering them is given by Michel Foucault (1977). He finds that underlying power relations in society manifest themselves in our forms of discourse and in the concepts that we may have inherited in language. What we approve of in our use of language reflects power relations hidden therein. It is not that knowledge is power, as Bacon thought, but that power is knowledge, as Foucault would have it. The study of science and technology, by professionals in recent times, constantly critiques and unmasks science, routinely showing that it is power speaking to truth. There are endless illustrations of Foucault's claim that knowledge claims are all about power relations at some deeper level. Foucault's examples are also illustrations of Popper's suggestion that endless confirmations are easy to find but pointless (Popper 1934).

Small bits of general knowledge do exist, contrary to the claim that science is all opinion, even if this knowledge is often ignored in the study of science and technology. Sometimes these bits are disquieting discoveries. One of them was made by Charles Darwin when he was a student of natural theology at Cambridge University. He wrote up a natural history of flora and fauna in the Galapagos Islands. Upon his return from a voyage there, he solved the puzzles that emanated from his copious notes. He found that the solution to his puzzles was not that an intelligent creator had organized the finches of the Galapagos Islands. Rather, the unique solution is that a species with many varieties in its offspring had spread to recently formed islands, where natural selection had determined which varieties had survived as new species. This model of heritable variation and natural selection was later extended to many other puzzles in biology. Those who do not wish to accept this bit of extended knowledge may do so in good faith on the ground that all generalities are hypothetical. All the best journals iterate this wisdom. But if we accept the inductive method of Bacon according to which we may sometimes discern a bit of reality more faithfully, then he had discerned one of them. If we accept it, then it seems that natural theology and many emblematic religious ideologies are in error.

A Fox in the Chicken Coop

The secular institutions of liberal democracy must try to remain scrupulously neutral between competing ideologies in order to retain our confidence. In a curious way, over the last 300 years, this role of secular institutions has gradually broken down, even as the secular professionals keep trying harder to remain neutral. One of these breakdowns in neutrality is in the discovery of how life evolves on earth, as we have noted. Another is due to the growing popularity of secularism itself.

We have seen a global trend toward growth in secular communities. Self-consciously formed, secular communities are often groups of atheists, agnostics, or humanists emulating a new religion. But there are also many informal secular communities whose members are not organized in this way. Many members may have a nominal other religious affiliation which does not define the scope of their activities. They do not practice their own religion. They do not necessarily socialize with others of the same religion and may trade, relax, or intermarry with members of nominally different religious groups. The rate of growth of these informal secular communities across the world has been astounding. A measure of their size is that they are roughly coextensive with urban living, even though this match is inexact. The accelerating pace of urbanization is a measure of the growth of informal secular communities.

In the past, religious groups may have occasionally felt threatened by conversion of members to other religions. Today, all religions face a common enemy: secularism. In 1950, about a third of the population on earth lived in cities, but now 55% do. By 2050, two thirds of the world's population will live in cities, if present trends continue. The only countries in which urbanization has stalled are where severe economic shock has set in, such as multi-year drought and famine. Ideological religions everywhere feel a need to fight this secular/urban growth.

It is not clear why urbanization is sweeping across the world. Part of the reason must have to do with the specialized jobs created in cities, where new and different skills can be learned and put to use. People can successfully exploit more of their own native abilities, where they may fail if opportunities were not available. For the ideologically religious, this must be a frightening spectacle: youth is drawn into secular communities without anyone converting them. Secularism was once an unbiased referee in a liberal democracy. It is now a formidable rival for the allegiance of all youth. It is a fox in the chicken coop. Not only does secularism form a rival community, but it also has a secular narrative that is formidable. Religious beliefs are now cast aside as opinions of outlying and undereducated groups, whereas secular opinions are well researched, thoughtfully disseminated, and backed up by formidable institutions.

Although the focus in this paper has been the anti-liberals of America, the fear of liberalism is everywhere in ideological religions. It is a growing concern where religions are being made emblematic for political purposes, witness Hinduism in India, Judaism in Israel, and Buddhism in Myanmar. In Islam, though their unifying rhetoric is often anti-Western, most of their policies are directed against liberalism. This is the whole point of the Saudi funding of madrassas or schools for the poor which teach only the Quran. The Saudis do not hate America, but they fear liberalism. Vicious attacks are conducted on secular schools that admit girls by the Afghan and Pakistani Taliban. Iranians shout "death to America," but it is their own population they wish to control, not America's. Consider, also, the atrocious actions by the Turkish government against its university employees and other intellectuals. This is fear of liberalism at its clearest. There is a reason why learning and education are under attack: these ideological religious groups are afraid of free thought.

A combination of the perfectly valid critique of Locke's theory of knowledge and the curious neglect of Bacon's successful experimental method of acquiring knowledge in science has had two unfortunate consequences. First, within the community of secular thinkers, there is no distinction made between things known and opinions that currently have institutional support. Both are cited with equal authority. This is perceived by religious groups as condescending. The secular experts are seen as much too full of themselves, trying to pass off their stuffy opinions as knowledge. Second, anti-liberals have concluded that *all* of their disagreement with the establishment is merely a difference in opinion. Nothing is right or wrong in and of itself. Any hypothesis may be defended come what may, as Willard Quine (1975) put it; and he taught philosophy at Harvard University, one of *those* institutions. If a fact needs to be denied, then deny it. We can make an accommodating adjustment of other thoughts later, when we have the time. In the meanwhile, we can call it an "anomaly." Kuhn said scientists normally do this, when they live with what they cannot even explain away (1962).

Moreover, as Feyerabend and Kuhn opine, different systems of thought have different lexicons. Facts cannot even be stated neutrally, so there can be no factual basis for comparing systems of thought. If scientific opinion reflects reality, then this is a reality constructed by the scientific group for a purpose. Let the elite university types have their opinion and their reality. The religious and marginalized still have their alternative, which, from a logical point of view, is none the worse for being "alt-." Whereas the "elite" have formidable institutional support, the marginalized have the "true news," the Gospel. What does not concur with it is construed as "fake news." Life is so simple. The victor will write our history, so there is only winning left, it seems. As Vince Lombardi of the Green Bay Packers is supposed to have said, exhorting his football team: "Winning is not everything" (pause), "it is the only thing." If such an attack on liberal democracy can be sustained, which is perhaps unlikely, and if liberal institutions are weakened enough, the currently marginalized will have won, achieving thereby conflict, cruelty, and bloodshed; which, according to the "elite," may be avoided.

It is a stretch of the imagination to suppose that after this supposed "victory" what remains will be kind to religious conservatism. How can urbanization be replaced by a new ruralization the world over? Will the cities and the specialized opportunities for our youth be allowed to dwindle? To achieve that victory, it is not only the "elite" that religions need to defeat but thereafter also the hopes, dreams, and aspirations of their own children.

References

Bacon, Francis. 2000 [1620]. *The New Organon*. Editor by Lise Jardine and Michael Silverthorne. Cambridge: Cambridge University Press.

Feyerabend, Paul. 1975. *Against Method: Outline of an Anarchistic Theory of Knowledge*. London and New York: New Left Books.

Foucault, Michel. 1977. *Discipline and Punish: The Birth of a Prison*. New York: Random House.

Horgan, John. 2017. Did Thomas Kuhn Help Elect Donald Trump? Filmmaker Errol Morris Claims Kuhn's Critique of Science Paved the Way for a Post-Truth Presidency. *Scientific American* (5/25/17).

Hume, David. 1993 [1777]. *An Enquiry Concerning Human Understanding*. Editor by Eric Steinberg. Indianapolis: Hackett.

Kuhn, Thomas S. 1962. *The Structure of Scientific Revolutions*. Chicago: University of Chicago Press.

Locke, John. 1965 [1690]. *An Essay Concerning Human Understanding*. Editor by John Yolton. London: Everyman Press.

Poincaré, Henri. 1905. *Science and Hypothesis*. Translated by W. J. Greenstreet. London: W. Scott.

Popper, Karl R. 1959 [1934]. *The Logic of Scientific Discovery*. London: Hutchinson.

Quine, Willard V.O. 1975. On Empirically Equivalent Systems of the World. *Erkenntnis* 9: 313–328.

CHAPTER 12

Ian C. Jarvie, Critical Rationalism and Methodological Individualism

Jeremy Shearmur

Introduction

In the course of this contribution, I will be making some criticisms of Ian C. Jarvie's ideas about methodological individualism. I would like to start by putting on record my appreciation for his powerful, dogged, and, in my view, for the most part highly successful development of critical rationalism in the social sciences. In particular, his *The Revolution in Anthropology* (1964) was especially important because of his engagement there with anthropological theory. By doing this so effectively, he seemed to me to offer those concerned with a Popperian approach a key lesson: engage with substantive issues, and methodological and epistemological issues in that context, rather than pursue methodology as if it were a self-contained discipline. Having said this, I have some reservations about the rest of my paper, for it itself takes up issues of a meta-level character! My excuse for doing this is that my space here is limited and because it seems to me that something has gone badly wrong with the *kinds* of things that critical rationalists have been trying to do in this area.

J. Shearmur (✉)
School of Philosophy, Australian National University, Canberra, ACT, Australia
e-mail: Jeremy.Shearmur@anu.edu.au

R. Sassower, N. Laor (eds.), *The Impact of Critical Rationalism*,
https://doi.org/10.1007/978-3-319-90826-7_12

Some Problems About Critical Rationalism and Methodological Individualism

Let me start by explaining why I think that there have been problems about critical rationalism and methodological individualism. At one level, everything is straightforward. Karl Popper favored methodological individualism. This he explicated by saying: "the task of social theory is to construct and to analyse our sociological models carefully in descriptive or nominalist terms, that is to say *in terms of individuals*, of their attitudes, expectations, relations etc." (1957, p. 126).[1] Popper was critical of approaches to the social sciences and history in which people set out to discover unconditional trends or sui generis cyclic patterns. He criticized approaches which took periodizations of history seriously, and he was also critical of what he called "psychologism"—the attempt to offer explanations in terms of psychology, rather than of individuals acting rationally on the basis of the logic of their situations (where their situations are understood as including statements as to what were taken to be their goals).

There were some complications about his view. In part, this was because of what could be called its mixed parentage: it was the offspring of Friedrich Hayek and Karl Marx (Shearmur 2017). When writing about his own ideas on the methodology of social science, Popper stressed the way in which he had, in effect, taken over and generalized Hayek's account of the methodology of microeconomics, so as to make it generally applicable to the social sciences and to historical explanation (Popper 1994, p. 154, 181). This also involved him situating it within his own ideas about epistemology and explanation. Popper stressed that one should see oneself as producing objective—and fallible—explanations, which, using the "rationality principle" in place of the universal laws of which one made use in the natural sciences, enabled one to deduce an explanation from initial conditions.

There are two contrasts with Hayek. First, Hayek's own ideas about methodology were distinctive, in the sense that he favored what he called a "compositive approach" (2010, Chap. 4). With this, explanations are built from things which we know directly concerning our own motivation (and the rules on the basis of which we classify things) which are taken also to be applicable to others. It also seems that Hayek broadly followed Ludwig von Mises (1960 [1933]) in his subjectivism: the idea that, pace Max Weber, all action could be seen as weakly rational. Popper could be seen as arguing that we should work with models of human action in which we tried to explain what was taking place as rational, given the situation as it was (and what we knew of the agent's knowledge, aims, etc.), rather than by seeing *all* action as rational. Popper went out of his way to stress that the ideas about rationality with which he was working could be *false* with respect to an acting agent (when, e.g., he was frustrated and tried to park a car in a space that he knew could not accommodate the car). When I was working as Popper's assistant and we were discussing "Models, Instruments and Truth" for inclusion in what became *The Myth of the Framework* (1994), I pressed on Popper a more Misesian view. This he resisted, but I was not at the time able to work out why. I am inclined now to explain this in terms of the fact that his discussion was concerned with the choice of the best and most fruitful and testable oversimplifying model rather than with a pursuit of a program which might send us on a quest for things which might be true, but not useful or testable, such as conjectures about whatever exactly might have been going through the mind of the person trying to park his car in a space that was too small.

All this has led some writers to suggest that Popper's "Models, Instruments and Truth" might be interpreted as setting out the approach of rational choice theory, the methodology of social science which has become increasingly dominant (Hedström et al. 1998). There is something to be said for this, especially in the context of this specific piece of Popper's. However, given the Hayekian background of Popper's ideas, the notion that we should identify such an approach with models using perfect knowledge seems to me odd, unless this is argued for in terms of their fruitfulness as models. But to take such a view seems to me to ignore a second key aspect of Popper's approach, and also ignore an important ambiguity with which Popper leaves us. For Popper's approach had two parents, and the second of these was Marx. There is a complication here: Popper's view of Marx is filtered through Popper's discussions with Karl Polanyi, which took the interpretation of Marx in an individualistic direction (Popper 1945, Chap. 3, Note 30). What is crucial here, and important for Popper's anti-psychologism, is that Popper follows Marx in stressing that people are to be understood as acting in *social* situations. But there is a contrast here with a rational choice approach, which typically directs us toward the preferences of agents. These, and the situations in which they are acting, are *implicitly* understood in historical and institutional terms. Yet these things do not figure explicitly in rational choice analysis.

Indeed, there seems to me a resulting ambiguity about Popperian methodological individualism. One might take it in one of two different directions. On the one side, we might be concerned with rational choice style analysis in which one works with a single, simple notion of human motivation. On the other, we might work with what could be called a more Weberian approach: one in which we work with a variety of models, chosen in the light of conjectures about people's motives and historical and institutional situations.[2] That there is a variety of such models does not mean that they have to be ad hoc: there could be explicit conjectures as to which model to choose and why, and once the subject matter is specified, such models may obviously be testable. (This view contrasts with Brennan and Buchanan 1981, pp. 155–166.) Compare the idea of working with a single model for the structure of all elements in chemistry as opposed to modeling different elements differently (provided that it is clearly identifiable what model is supposed to apply to what and that what is being claimed about any specific element is testable). In my view, there is no reason why we should not work with what I call the Weberian view, although we should be interested if the proponents of a more unitary approach can claim that their view offers *better* explanations.

In the light of this point, it seems to me that we might also reconsider Popper's (and also Weber's) anti-psychologism. Within "behavioral economics," some interesting ideas have been developed on the basis of work which is both empirical and theoretical in its character, and challenges have been advanced to economists' typical rational choice approaches (Thaler 2015; Latsis 1972; Hausman 1992). On the face of it, if we can admit Weberian approaches when they are properly testable, then we should also admit approaches which are psychologistic in the sense of this behavioral tradition and accept them as part of our tentative knowledge (provided their testability and performance are better than that of standard neoclassical models). Clearly, if neoclassical approaches are understood as oversimplifying models, then simply to point out that they are not true does not, in itself, settle the matter; one needs to know also that if their assumptions are modified along the lines suggested by behavioral work, it will offer us interesting, testable, and better models.

The Status of Methodological Individualism

One oddity about Popper's advocacy of "methodological individualism" relates to its status. As I have argued elsewhere (Shearmur 2017), it seems important to bear in mind, when approaching Popper's work, that his ideas about what was open to rational assessment when he wrote *Die beiden Grundprobleme* and *Logik der Forschung* (cf. 1959) were limited in their scope. While Popper was conscious of holding metaphysical theories, and certainly did not think that metaphysics was meaningless, one might say that in this early work what was taken to be open to rational scrutiny was mathematics and other formal claims, empirical claims, and also methodological claims—the latter, however, to be assessed in the light of chosen aims. In addition, a theme of those two first books was that metaphysical claims should be translated into and assessed as methodology. While in Popper's *Open Society* (1945, Chap. 24, Sect. I), one starts to find some ideas about a "rationalist attitude" developed by way of analogy with his ideas about science, the bulk of the book is written in the same mode as his earlier work. It is only later, once he gets to working on *The Postscript*, that we find him elaborating a theory of how we might evaluate metaphysical claims.

I wish to highlight here that Popper's methodological individualism is explicitly a methodological theory. Yet Popper does not treat it in the same way as he does his methodological ideas in, say, *The Logic of Scientific Discovery* (1959), that is, as something that is to be evaluated instrumentally in relation to our general ideas about the aims of science. Indeed, one might say that it is introduced without our being told how it is to be evaluated. The reader might ask: Just why should we go for methodological individualism as opposed to an alternative approach? And, indeed, with just what is it to be contrasted? The main candidates here seem to be the psychologism of which Popper is critical as leading to a kind of regress (a criticism which would not hold good of the behavioral approaches discussed above) or the various kinds of historicism and holism of which Popper is also critical. Popper is cagy about its status, while there is also a danger that if its content is to be understood as simply a negation of the other things to which I have referred, it is too vague to be of much use in the social sciences (or too weak to rule out anything).

It is striking that while Noretta Koertge (1974) discusses Popper's ideas about the methodology of social science as a research program, Popper's own work on methodological individualism was not offered as a

Popperian research program. I am not complaining that Popper did not develop his ideas in such a form himself, just because the development and articulation of his ideas about research programs and their role came after the bulk of his work on methodological individualism, and later he had many other more pressing things to do than to undertake work about the status of his ideas concerning the social sciences. But one might usefully contrast Popper's ideas about research programs when he did develop them with his treatment of methodological individualism when he introduced it.[3] When Popper wrote about metaphysical research programs, he discussed the way in which they might be directed at a problem situation in science. They offer an account of what the world is like, and of the plausibility of such a view (where the idea of plausibility has to be understood in a way compatible with Popperian anti-justificationism). In addition, they then suggested how one might approach problems in the field toward which the ideas were directed in a manner which was compatible with the research program. But methodological individualism is methodological rather than metaphysical (in Popper's later approach one might expect a *metaphysical* theory, and then an account of what methodological approach it would be sensible to make use of, should the metaphysical approach be correct). In Popper one also does not really have the theory developed in relation to a problem situation in any of the social sciences. In this respect, Jarvie (1964) seems to me a real improvement, just because it was concerned with theoretical approaches in a specific discipline.

Watkins, Jarvie, and the Moral Thrust of Methodological Individualism

I argued elsewhere (Shearmur 1996) that Robert Nozick had made an important criticism of methodological individualism. This was that some social phenomena might usefully be understood as the products of interactions between individuals and what he called "filter effects." These are well known to us in the physical world. For example, the floor of the passage to a room might only support weights of 150 pounds, which then means (assuming that we reached the room via the passage) that those who are in the room will be those who (1) wished to enter it or were forced to come along, by others, and (2) weighed less than 150 pounds. It is also plausible to think of aspects of the social world as operating on such a basis. From Adam Smith through Marx to Milton Friedman (and in many more recent "evolutionary" approaches in the social sciences), we have been offered various accounts of the ways in which social phenomena may be the products of such effects (Shearmur 1983). Typically, the filtering mechanisms are the products of human action but not of human design, which take on a thing-like character and which may act back upon us. They have to be sustained by ongoing human action in various social situations. What is crucial here is that what does the filtering may have the character of emergent products, relative to the actions which give rise to them. It is important to appreciate that what is involved in this is not the idea—which poses no problem for methodological individualism—that individuals act in a social setting, in the sense that roles, traditions, judgments about social situations, and so on will be internalized into people's understanding of the situations in which they act. It is, rather, the idea that emergent social entities can have a thing-like character and interact with our actions, much as do physical objects. In addition, we may then need to investigate how individuals, in the face of such things, weigh the options open to them, if, for example, they are not able to accomplish what they had initially hoped to achieve because they encounter such constraints.

This is important for the methodology that we adopt in the social sciences. While what is involved will differ, depending on the specific theories about the characteristics of such social products that we may wish to adopt, we may need to operate both at the level of the acting individual and also at that of the emergent filter mechanism if we are to make sense of what is going on. It is, however, something that clearly contrasts with what Hayek wrote about the "compositive" method, which is supposed simply to start from individual action. And it seems to me to call for clarification from those in the Popperian tradition as to just what they are advocating. Jarvie, somewhat to my surprise, reacted to this problem by offering a minimalist reading of methodological individualism: "methodological individualism consists mainly in the denial of aims to social wholes" (2016, p. 313, n. 30). While this may be true enough, it seems to me unhelpful at the level of methodology, for it is difficult to imagine anything that is actually being advocated in the social sciences which is incompatible with such a view. And we know, on broad Popperian grounds, that being compatible with just about everything is something problematic rather than good, as it suggests that we are dealing with something with little content.

There was, however, a second thrust to Jarvie's response. He argued that a key concern of methodological individualism was in fact moral: that it should be understood as repudiating ideas about collective aims and historical inevitability, which would rob the individual of agency: “If we grant that collectives act in ways that are quite irreducible to the actions of individuals, then a certain fatalism is permitted: we are creatures of forces greater than ourselves and these forces will drive towards their outcome no matter what we do” (Ibid., p. 303). This, for reasons which I shall explain, surprised me. It is not that I would disagree with a scholar’s being motivated to develop such an approach to the social sciences because of moral concerns. But it would seem to me to make little sense as providing a rationale for *methodological* individualism. Before I move to criticize such ideas, I should stress, however, that Jarvie is by no means alone in holding such a view. John Watkins, in correspondence with Arthur Danto in 1962,[4] made a very similar statement concerning what he considered to be the real point at issue concerning methodological individualism.

What is wrong with this as an explication of methodological individualism? It is that methodological individualism, as its name indicates, is a *methodological* theory. (Jarvie (Ibid., 304) goes out of his way to argue that Popper was not an ontological individualist.) What is needed to resolve the problem that concerned Jarvie and Watkins would seem to me in part the criticism of theories at odds with ideas about such individual freedom but also to be the advancing of a substantive scientific, metaphysical, or ontological theory about the role of human agency in history. We would need a plausible account that the world is such that individual actors have the opportunity for moral agency of whatever kind is needed to make Jarvie’s and Watkins’ moral case. I am sympathetic to such an enterprise. But what is called for is a full-fledged theory of human autonomy, and then a discussion of such matters as whether, if people have the relevant kind of metaphysical freedom, this would be sufficient for our purposes if, for example, the costs of acting on the basis of such capacities would be very high. I think that Popper has offered us important ideas about human autonomy (which typically but still strangely seem to me to have been neglected by those writing on the topic). What does not and, I would suggest, *cannot* help here is a theory of purely *methodological* individualism. The main weight of the argument has to be carried by substantive rather than methodological theories. I am *not* arguing here for atomistic individualism, just that a *methodological* thesis seems to be the wrong way to go if our concern is with human freedom or autonomy.

The Task Before the Popperian Methodological Individualist

Let me break a lance for some of the issues related to how methodological individualism needs to be developed here by way of engagement with concrete problem situations in the various social sciences. It seems that three different kinds of ideas need to be brought together. The first—which is something that Popper disavowed, but for which in light of his *later* work in philosophy there would seem no reason to abjure—is to offer a tentative account of what one takes to be the relevant ontology here. From a Popperian perspective, ideas which would contribute to this are already available: that we are biological creatures who have a limited degree of autonomy (which relates importantly to ideas about intersubjective critical scrutiny and its internalization; on the latter compare Vygotsky (1986)). We act in social situations. In this context, the emphasis that Adam Smith and French theorists of that period gave to our mode of subsistence is of particular importance. But so are our *ideas.* These involve a number of things: our intellectual understanding of the world, physical and social, and various rules, practices, and standards into which we are born and into which we are socialized. Also important are traditions and habits, of which we may not be aware in any self-conscious way.

Second, there is the fact that our actions—which one might understand as taking place on the basis of *all* of the above, and as being subjectively rational in the weak Misesian sense of being what we take to be the best option at the time—have consequences, including the production of unintended consequences. These may—as with linguistic innovations or contributions to new paths—affect how people later do things: path dependency may then become significant! In addition, these actions may create emergent thing-like objects which interact with and constrain (or facilitate) subsequent actions. All of this indicates why an approach in the social sciences after the fashion of Peter Winch (1958) or John Searle (1996), which stresses ourselves as rule-governed, is inadequate. For the unintended consequences of our actions, by which we are affected, may be things which we don't understand or which we may misunderstand. Clearly, we will act on the basis of our subjective understanding of our situation. But we may misidentify the character of what we are dealing with, or our ideas and practices may incorporate a theory about their character which is mistaken (as when, e.g., people invest on the basis of a theory about the character of the trade cycle, which is, in fact, incorrect).

Third, there is the issue of how all of this relates to the problem situations in the various social sciences. I clearly cannot offer a survey here of any depth, but a couple of issues might be worth a comment.

The first set of issues relates to economics. A major problem would seem to me to be the shift away from theory to purely technical problem-solving, where this is also unconnected with real-world economic issues. Here, a Popperian shift to historical-based appraisal (the conditions for the acceptability of a theory should include its offering a better resolution than did its predecessors of the problems to which they were addressed) would be invaluable. A critical history of economics could then play a really important role. Further, should we use a single model of human action (maximization on the basis of preferences), and what should we take these preferences to be? It seems to me that there should be no objection to making use of specific models (the content of which might be usefully informed by empirical work in sociology, psychology, or experimental economics), provided that these are testable and generate better explanations of phenomena and better resolutions of problems than do standard models. Greater descriptive realism may be suggestive; but it will not necessarily be helpful. For example, preference reversal under certain circumstances may occur (Hausman 1992), but whether it would be useful to change our models to reflect this is quite another matter. Given that Popper's ideas in this field took off from Hayek's ideas about what he called the "compositive" method, three points might be worth noting.

1. Our method should be termed *resoluto-compositive*: we should build into our approach the idea that we typically start from some macro-level problem (e.g., patterns which, as individualists, we take to be the products of human action in social situations and try to explain in such terms). Hayek, in taking about his own approach, referred to Carl Menger's (1963) reaction to those among his contemporaries who made much of organic analogies in the social sciences. Menger argued that these should not be dismissed but rather explained individualistically.
2. Hayek's approach faces the problem: At what level of aggregation (and simplification) should we work when building our models? This is not resolved should we accept his method. If we start from the actions of concrete individual agents, then it is left completely open as to how we move from these to the content of our social theories. These typically don't refer to specific individuals and offer

very general and often stylized ways of representing their actions and may group aspects of them together in terms which would not have been used by those individuals themselves (e.g., the idea of "liquidity preference"). In my view, how we represent actions should be guided by our theories about the character of the phenomena we are trying to explain. There should be an interplay here between what is going on at the macro and the individual levels, with the description of action at an individual level in many respects being an artifact of our explanatory enterprise, rather than involving a procedure which is simply compositive.

3. While we may admit to the significance of commonsense knowledge, including about individual human motivation, any attempt to use this as the foundation of an approach in economics hits a problem raised by Pierre Duhem (see Shearmur 1991). Many different ways are open to us as to how such commonsense ideas are to be represented technically, while our choice between these representations is underdetermined by our knowledge at a commonsense level.

The second set of issues relates to anthropology. As I have indicated, Jarvie's (1964) contribution was of real significance. But several issues seem to me still to be pressing.

1. Functionalism in anthropology. There seems to be a contrast between the way in which this concept is widely used by those engaged in fieldwork, while it has come in for considerable criticism not just in Jarvie's work but also in the Jon Elster (1982) and Gerald Cohen (1982) debate. On the face of it, functionalism that can be explained in terms of something analogous to natural selection is unexceptionable. Cohen argued that we should retain our knowledge of functional relationships for which we have strong empirical evidence but for which we do not, currently, have such an explanation. I would suggest taking a similar line toward structuralism. There was a fad for explanations in social affairs which claimed to disclose the existence of an underlying structure to myths, folk practices, and so on. Two problems arise here: first, it is not clear how strong the evidence for such explanations was, while what was supposed to underlie such things never seemed to me to make much sense. Similar ideas were advanced (Foucault 1972; cf. Part 4 Chap. 4) in terms which seemed to explicitly preclude any kind of independent assessment, and which as a result seem to me of little value.

My suggestion would be that insofar as interesting patterns are claimed to exist, then unless testable theories can be advanced to offer sui generis explanations for them, we should take a Mengerian approach of explaining them individualistically. At the same time, those who make use of such ideas (e.g., in terms of relating people's explanatory ideas to their social structure in the Durkheimian tradition) should be encouraged to explicate them in as powerful and general a form as possible.

2. We should do much more by way of participant observation to investigate what people's culture and beliefs actually are. Such work (which typically involves study over a long period of time) can be invaluable in correcting our presuppositions. (See, e.g., work on fundamentalism in the US, such as Ault 2004 and Peshkin 1986.) At the same time, it seems to me of the greatest importance that there is an interplay between such work and empirical studies which test whether or not the kinds of relationships that are discovered in particular cases hold good more generally.
3. We should be careful lest we overintellectualize an individualistic approach. Jarvie and Agassi's (1967) work on the rationality of magic, while very important and I believe for the most part correct, seems to me to run into the danger of doing this. They seem to suggest that to understand people's actions, we should seek to understand the view of the world behind them. One might caricature their approach as hunt the cosmology. But while this may be both important and fruitful, there seem to me a couple of problems about it. The first is that, as Ault's (2004) study of a fundamentalist community makes clear, actual practices may be far from an application of people's avowed beliefs. One had in this case people who avowed a form of principled Biblical fundamentalism but whose actual community practices seem to be closer to a form of situational ethics! The second is that if we accept that much in our society and behavior is traditional, and also if the unintended consequences of human actions play a significant role in our society, then much that we do will simply be a matter of doing what is customary rather than anything that has any kind of strong rationale behind it. Our actions are weakly rational in a Misesian sense. But there may be no deeper explanation for them, other than a historical one which, *if* it can be reconstructed, may serve to explain how it was that people of this kind in this particular situation ended up behaving in this kind of

way.[5] It is worth bearing in mind the way in which intelligent young people are regularly scandalized when they discover that people don't behave on the basis of the abstract principles that they profess. Some of the unintended consequences of our actions (e.g., the occurrence of economic depression) may baffle us, such that we resort to poor-quality social theory (e.g., conspiracy theories) in order to explain what is going on. It is also striking that some religious practices seem to develop by analogy, in circumstances in which it is not clear that any rationalistic explanation could fully make sense of what is taking place.[6]

Rather than continuing this survey, I will conclude by simply restating my suggestion that what is needed is a dual-pronged development of Popperian methodological individualism, both in terms of strengthening its theoretical ideas and exploring concrete problem situations in the social sciences. I hope that this is a task in which younger critical rationalists will follow Jarvie's important lead.

Acknowledgment I would like to thank Francesco Di Iorio for his comments on an earlier version of this paper and Raphael Sassower for many useful suggestions. I am only sorry that considerations of space precluded my responding to all the interesting problems which Di Iorio posed.

Notes

1. See Popper's "Models, Instruments and Truth" and "A Pluralist Approach to the Philosophy of History" (1994); see also "Historical Explanation" (2008) and (1972, Chaps. 3 and 4).
2. My concern here is with the more individualistic side of Weber's approach, as found in his explicitly methodological writings (which may be contrasted with aspects of his approach to politics, which were at times more Marxian in their character); see on the latter Beetham (1974).
3. I have no objection to Koertge's approach to Popper's views—her discussion ranges far beyond methodological individualism. My point, rather, is that when Popper introduced "methodological individualism," it did not fit the ideas about "methodological rules" that he held at the time, while his ideas about "metaphysical research programs" were only articulated later and which also don't fit his treatment of methodological individualism. Popper had long held that some metaphysical ideas played an important role in the development of science. But his ideas about "metaphysical research

programs" were developed first in his lectures in 1949 and set most fully in his *Unended Quest* (Popper 1976) and the *Postscript*.

4. See Watkins to Danto, 11 January 1962; L.S.E. Archive J. W. N. Watkins, Box 14, Folder 136.
5. Compare the tradition, in some Western societies, of throwing salt over one's left shoulder if it is accidentally spilled on the table or the custom, in England, of wearing paper hats from Christmas crackers, during Christmas dinner. Such practices have historical origins; but these are typically not known to those who behave in this manner.
6. An interesting case is the development of the burning of (fake) paper money for the sake of the dead, in popular Chinese religious traditions, into the burning of paper versions of televisions, cars, and even Viagra. It seems that the burning of paper money may have had its origins in an indulgence-like belief that it could assist people in avoiding penalties in hell, prior to their reincarnation. The extension of this seems to be a matter of analogy, built on a misunderstanding of the original rationale for burning the paper money and to a picture of hell that makes no sense, although one might suspect that the *content* of TV programs in hell might be very much like the lower range of channels on satellite TV.

References

Ault, James A. 2004. *Spirit and Flesh*. New York: Knopf.

Beetham, David. 1974. *Max Weber and the Theory of Modern Politics*. London: Allen & Unwin.

Brennan, Geoffrey, and James Buchanan. 1981. The Normative Purpose of Economic 'Science'. *International Review of Law and Economics* I: 155–166.

Cohen, Gerald A. 1982. Reply to Elster on 'Marxism, Functionalism, and Game Theory'. *Theory and Society* 11 (4): 483–495.

Elster, Jon. 1982. The Case for Methodological Individualism. *Theory and Society* 11 (4): 453–482.

Foucault, Michel. 1972. *The Archaeology of Knowledge*. London: Tavistock.

Hausman, Daniel. 1992. *The Inexact and Separate Science of Economics*. Cambridge: Cambridge University Press.

Hayek, Friedrich A. 2010. *Studies on the Abuse & Decline of Reason*. Edited by Bruce Caldwell. Chicago: University of Chicago Press.

Hedström, Peter, Richard Swedberg, and Lars Udéhn. 1998. Popper's Situational Analysis and Contemporary Sociology. *Philosophy of the Social Sciences* 28: 339–364.

Jarvie, Ian C. 1964. *The Revolution in Anthropology*. London: Routledge.

———. 2016. Popper's Philosophy and the Methodology of Social Science. In *The Cambridge Companion to Popper*, ed. Jeremy Shearmur and Geoffrey Stokes, 284–317. Cambridge: Cambridge University Press.

Jarvie, Ian C., and Joseph Agassi. 1967. The Problem of the Rationality of Magic. *The British Journal of Sociology* 18: 55–74.

Koertge, Noretta. 1974. Popper's Metaphysical Research Program for the Human Sciences. *Inquiry* 18: 437–462.

Latsis, Spiro. 1972. Situational Determinism in Economics. *British Journal for the Philosophy of Science* 23: 207–245.

Menger, Carl. 1963 [1883]. *Problems of Economics and Sociology*. Urbana: University of Illinois Press.

von Mises, Ludwig. 1960. *Epistemological Problems of Economics*. Princeton: Van Nostrand.

Peshkin, Alan. 1986. *God's Choice*. Chicago: University of Chicago Press.

Popper, Karl R. 1945. *The Open Society*. London: Routledge.

———. 1957. *The Poverty of Historicism*. London: Routledge.

———. 1959 [1934]. *The Logic of Scientific Discovery*. London: Hutchinson.

———. 1972. *Objective Knowledge*. Oxford: Clarendon Press.

———. 1976. *Unended Quest: An Intellectual Autobiography*. London: Fontana.

———. 1994. *The Myth of the Framework*. London: Routledge.

———. 2008. *After the Open Society*. London: Routledge.

Searle, John. 1996. *The Construction of Social Reality*. London: Penguin.

Shearmur, Jeremy. 1983. Subjectivism, Falsification and Positive Economics. In *Beyond Positive Economics?* ed. J. Wiseman, 65–86. London: Macmillan.

———. 1991. Common Sense and the Foundations of Economic Theory: Duhem vs. Robbins. *Philosophy of the Social Sciences* 21 (1): 64–71.

———. 1996. *The Political Thought of Karl Popper*. London: Routledge.

———. 2017. Popper's Influence on the Social Sciences. In *The Routledge Companion to Philosophy of Social Science*, ed. L. McIntyre and A. Rosenberg. London and New York: Routledge.

Thaler, Richard. 2015. *Misbehaving: The Making of Behavioral Economics*. London: Allen & Unwin.

Vygotsky, Lev. 1986. *Thought and Language*. Cambridge: MIT Press.

Winch, Peter. 1958. *The Idea of a Social Science*. London: Routledge.

CHAPTER 13

Jarvie on Rationality and Cultural Relativism

Kei Yoshida

Introduction

Ian C. Jarvie is widely known for his contribution to the "rationality debate" in philosophy of the social sciences. With Joseph Agassi, Jarvie argues that rationality can be seen as a matter of degree. Furthermore, he tries to show that rationality is not a matter of individual mentality, but one of action and contexts. By doing so, he emphasizes the role of contexts or social institutions. From such a perspective, he has severely criticized cultural relativists. In this chapter, I shall critically examine his view of rationality and rebuttal of cultural relativism. First, I shall discuss his view of rationality while paying attention to its relation to Karl Popper's critical rationalism. The former can be regarded as a development or refinement of the latter. For this reason, it is necessary for us to examine how critical rationalism developed. Second, I shall scrutinize Jarvie's criticism of cultural relativism. Although Jarvie's debate with Peter Winch in the early 1970s is famous in philosophy of the social sciences, here I shall investigate how Jarvie criticizes cultural relativism after his confrontation with Winch. Jarvie proposes a middle way to avoid ethnocentrism and cultural relativism by examining Melville Herskovits's cultural relativism

K. Yoshida (✉)
Faculty of Social Sciences, Waseda University, Tokyo, Japan
e-mail: kei.yoshida@waseda.jp

R. Sassower, N. Laor (eds.), *The Impact of Critical Rationalism*,
https://doi.org/10.1007/978-3-319-90826-7_13

and its relation to the American Anthropological Association's "Statement on Human Rights." Third, I shall examine Jarvie's social and institutional view of rationality by discussing his ideas of weak absolutism, workshop rationality, and social institution. He argues that a social and institutional character of rationality is necessary for the growth of knowledge: the unintended consequence of human actions is crucial to social institutions. The relation between his view and Francesco Guala's recent work on social institutions will also be discussed.

Contextualizing Jarvie's View of Rationality in the Development of Critical Rationalism

The problem of rationality has been one of Jarvie's major problems throughout his career. There are at least two reasons for this. First, as a former graduate student of Popper's, he is a critical rationalist; however, being a critical rationalist is common to most, if not all, of Popper's former graduate students and thus is not unique to Jarvie. The second reason is that Jarvie obtained his bachelor's degree in social anthropology from the London School of Economics and Political Science (Jarvie 1984, p. 123). As far as I know, Jarvie is the only student who has an official training in social anthropology and a great interest in the social sciences among Popper's former graduate students. Even Jeremy Shearmur, who works on the political philosophy of Friedrich Hayek and Popper, admits this point (Shearmur 2017, pp. 61–62). As a result, this puts Jarvie in a unique place among other critical rationalists and suggests his original contribution to the problem of rationality. Jarvie's significant contribution is his application of critical rationalism to anthropology, where a cross-cultural comparison of customs, habits, and values is one of its prime concerns. By doing so, he has shed light on the problem of explaining different cultures and proposed a different solution from both ethnocentrism and cultural relativism. Jarvie uses the term "absolutism," too. He seems to regard absolutism contrary of relativism and to take its anthropological alternative as ethnocentrism (Jarvie 2007, p. 557). In his endeavor to propose a middle way between ethnocentrism and cultural relativism, Jarvie is not alone. He has a lifelong fellow traveler, Agassi. Jarvie and Agassi argue that rationality can be seen as a matter of degree. In proposing such a view, Jarvie is influenced by the development of critical rationalism. Thus, it is necessary for us to know it and situate Jarvie in its context.

But in contextualizing Jarvie's view of rationality in the development of critical rationalism (for details, see Yoshida 2014, pp. 6–10; 26–29), where should we begin? Many commentators on Popper find this to be a good starting point: "*I may be wrong and you may be right, and by an effort, we may get nearer to the truth*" (Popper 1945, Vol. 2, p. 213; italics in the original). This is a fundamental attitude of rationalism that requires us to be open to criticism and learn from errors. But Popper argues that the adoption of the rationalist attitude is based on "an irrational *faith in reason*" and that "[r]ationalism is therefore far from comprehensive or self-contained" (Ibid., p. 218; italics in the original). William Bartley, another student of Popper's, finds this problematic and develops his own version of critical rationalism to overcome Popper's faith in reason. He calls it "comprehensively critical rationalism" or "pancritical rationalism" (Bartley 1984). There was a huge debate about whether Bartley's view can overcome Popper's problem. Jarvie participated in it. Although this debate itself is interesting, let me keep it aside and go back to the development of critical rationalism.

The crux of Bartley's argument is that by differentiating between justification and criticism and abandoning the former, we have a possibility of non-justificational criticism; criticism does not need justification. What does this claim mean? It would be helpful to look over traditional rationalism which can be characterized as trying to find a firm foundation for knowledge and science. René Descartes's famous proposition "*I think, therefore I am*" is regarded as its typical example (Descartes 1911, p. 101; italics in the original). According to Bartley, any position seeking for a firm foundation can be called justificationist. He claims that justificationism cannot escape from what he calls "the dilemma of ultimate commitment," namely, the dilemma between infinite regress and dogmatic commitment (Bartley 1984, pp. 72–73). When searching for a firm foundation, justificationists must respond to endless inquiries of whether what they found is really a firm foundation. They need to endlessly justify their claim. This is infinite regress. If they want to avoid such infinite regress, they are forced to dogmatically claim that what they found is a firm foundation. This is dogmatic commitment. Bartley claims that justificationists must face the dilemma because they tacitly identify justification with criticism. Thus, Bartley argues for the need of non-justificational criticism (Yoshida 2014, pp. 6–7).

From such a non-justificational perspective, Bartley tries to modify Popper's critical rationalism. Popper demarcates between science and non-science by using a falsifiability criterion. He argues that if a theory is not falsifiable, then it is not scientific. He calls any theory that is immune to criticism, and thus unfalsifiable, a reinforced dogmatism (Popper 1940, p. 417). Popper regards such theories as Marxism and Freudian psychoanalysis as typical examples of reinforced dogmatism. According to Bartley, the demarcation between the rational and the irrational or between the critical and the uncritical is more important than Popper's demarcation between the scientific and the nonscientific (Bartley 1968, p. 64). We can find many examples of reinforced dogmatism not only in science but also in ethics, metaphysics, and so on. It does not make sense to limit the scope of critical rationalism to the field of science. Bartley modifies critical rationalism by making it applicable to fields other than science (Yoshida 2014, p. 7).

Bartley's view has some limitations. Instead of Popper's demarcation, Bartley proposes his own demarcation between the rational and the irrational. This shows that Bartley presupposes a strict distinction between the two, as Popper thinks about the demarcation between science and nonscience. Jarvie and Agassi try to blur such a strict distinction by introducing a new idea of rationality. According to them, rationality is better seen as a matter of degree. Nothing can be regarded as being completely rational or irrational. Even if something looks irrational to us, it contains some part of rationality. Jarvie and Agassi modify Bartley's view.

By regarding rationality as a matter of degree, what do Jarvie and Agassi mean? They argue that rationality can be distinguished into three levels. Rationality 1 relates to "the goal-directed action of an agent with given aims and circumstances" including his/her knowledge and opinions (Jarvie and Agassi 1979, pp. 353–354). As examples of this kind, Jarvie and Agassi refer to the magician's use of incantation, the traditional doctor's bloodletting, and the modern doctor's use of antibiotics for the treatment of illness. These examples differ from each other in detail, but they attempt to solve the problem of treating illness by their own methods. For this reason, they can be taken as satisfying one level of rationality.

Rationality 2 includes "the element of rational thinking or thinking which obeys some set of explicit rules" (Ibid., p. 354). We cannot find this level of rationality in magic in general, although it may be applied to some particular parts of magical thinking. Jarvie and Agassi refer to Edward Evans-Pritchard's observation that although the magician may be critical or rational about specific details of magical thinking, he/she cannot be critical or rational about magic in general. In this sense, "magic in general is pseudo-rational because it lacks [articulate] standards of rational thinking" (Ibid., p. 354).

Rationality 3 is "that of goal-directed action ([R]ationality 1) subject to thinking that conforms to not just some rule or standard of rational thinking ([R]ationality 2), but the best available rules for, the highest standards of, rational thinking" (Ibid., p. 354; see also Jarvie 2005, pp. 475–476). This is how Jarvie and Agassi argue for the possibility of degrees of rationality (for details, see Yoshida 2014, pp. 26–27).

Here we can discern Popper's influence on Jarvie and Agassi's distinction between Rationalities 2 and 3. Popper contends that science differs from myth, because the former has the second-order tradition of criticizing the latter: a critical attitude toward one's own tradition. By taking a critical attitude toward a myth, the myth changes and transforms into providing a better explanation about the world (Popper 1949, pp. 44–45). From such a claim, we can say that magic in general belongs to Rationality 2, because it does not have the second-order tradition of critical discussion. By contrast, science can be regarded as belonging to Rationality 3, because it has the second-order tradition. But this does not mean that science is always rational, as it is often claimed. To achieve Rationality 3, it is important to satisfy the highest standards of rational thinking, to be critical of one's own view. But if we recall Thomas Kuhn's view of paradigm and normal science, we can see that scientists may adhere to their favorite theories (Kuhn 1996). This means that they may not be sufficiently critical of their views. But there are social institutions where other scientists can criticize the reigning paradigm. The reason science can be seen as more rational than magic in general depends not only on people's attitudes toward their own views but also on the institutional character of science (Yoshida 2014, pp. 27–28). This institutional view applies not only to science but also to rationality. We have so far contextualized Jarvie's view of rationality in the development of critical rationalism. With Agassi, Jarvie has argued for the possibility of degrees of rationality. But how does he criticize cultural relativism?

Jarvie's Rebuttal of Cultural Relativism

Among his rebuttals of cultural relativism, Jarvie's criticism of Winch is widely known in philosophy of the social sciences (for details, see Yoshida 2014, pp. 15–31). Winch criticizes Evans-Pritchard and proposes a different approach to understanding other cultures (Winch 1958, 1964). According to him, the social sciences differ from the natural sciences insofar as the former must internally understand social phenomena, while the latter causally explain natural phenomena (without paying attention to their internal states). From such a perspective, Winch argues that different criteria of rationality correspond to different aspects in our daily lives, and the social sciences must understand them, respectively. Scientific practice is just part of our daily lives and does not exhaust all possibilities of human rationality (Winch 1970, p. 250). For this reason, there are different rationalities, scientific, religious, aesthetic, and so on. Though Winch denies that he is a relativist, he is often criticized as such. In his debate with Winch, Jarvie claims that Winch is not a relativist in the complete sense of the word but still gives in too much to relativism (Jarvie 1970, p. 232). Jarvie writes that there is the minimum standard of shared rationality: "[T]here is something like a community of rationality shared by all men, but recognized or fostered by different societies in varying degrees (none being perfect)" (Ibid., p. 238). Jarvie thus rejects Winch's view. We can easily discern that there is a relation between Jarvie's criticism of Winch and his idea of degrees of rationality.

The debate between Jarvie and Winch is intriguing, but it has been discussed for many years since the early 1970s. Thus, I shall no longer repeat it here. After his confrontation with Winch, how does Jarvie criticize cultural relativism? In answering this question, Jarvie's several works on relativism would be helpful (Jarvie 1983, 1984, 2007). In these works, Jarvie refers to Melville Herskovits's cultural relativism which is contrasted with ethnocentrism. Ethnocentrism means that "one's own way of life is to be preferred to all others" (Herskovits 1972, p. 21). Herskovits describes the principle of cultural relativism as follows:

> *Judgments are based on experience, and experience is interpreted by each individual in terms of his own enculturation.* Those who hold for the existence of fixed values will find materials in other societies that necessitate a re-investigation of their assumptions.... Even the facts of the physical world are discerned through the enculturative screen, so that the perception of time, distance, weight, size, and other "realities" is mediated by the conventions of any given group. (Ibid., p. 15; italics in the original)

Herskovits emphasizes here that enculturation influences the development of our judgments and the perceptions of physical realities.

From such a relativist perspective, Herskovits participated in drafting a document submitted to the Commission on Human Rights of the United Nations which was preparing for the Universal Declaration of Human

Rights; the American Anthropological Association drafted the document to provide some input to it (AAA 1947). Although this document was submitted under the name of the Executive Board of the AAA, it is said that Herskovits primarily wrote it (Bidney 1953, p. 693; Engle 2001, p. 539). The statement aims at resolving the problem of how the Declaration is applicable to all human beings, not just to people who accept Western values (AAA 1947, p. 539). This document is notorious and embarrassing, because it is regarded as espousing cultural relativism whose origin can be traced back to Herskovits and his former supervisor Franz Boas. We can discern Herskovits's influence on the statement in the following passage:

> If we begin, as we must, with the individual, we find that from the moment of his birth not only his behavior, but his very thought, his hopes, aspirations, the moral values which direct his action and justify and give meaning to his life in his own eyes and those of his fellows, are shaped by the body of custom of the group of which he becomes a member.... Yet if the essence of the Declaration is to be, as it must, a statement in which the right of the individual to develop his personality to the fullest is to be stressed, then this must be based on a recognition of the fact the personality of the individual can develop only in terms of the culture of his society. (Ibid., pp. 539–540)

The above passage says that enculturation plays an important role in developing individuals in their societies. They cannot fully develop outside their own societies. It is not difficult to see how similar this is to Herskovits's principle of cultural relativism. Jarvie argues that there are several problems with the above passage. First, it implies that individuals with cross-cultural backgrounds cannot develop fully. Second, it also implies that exiting from one's society in the process of development is harmful to an individual. Jarvie points out that since the AAA represents US anthropologists, where most people are the descendants of immigrants, the statement is confused (Jarvie 2007, pp. 571–572). Furthermore, Jarvie closely examines whether the AAA statement makes nonrelative moral and cognitive claims while arguing for cultural relativism. According to him, while the statement criticizes Western economic expansion to and control over non-Western countries, it endorses respecting cultural diversity. Jarvie investigates what is wrong with this and focuses on the following principles in the AAA statement:

1. *The individual realizes his personality through his culture, hence respect for individual differences entails a respect for cultural differences.*
2. *Respect for differences between cultures is validated by the scientific fact that no technique of qualitatively evaluating cultures has been discovered.*
3. *Standards and values are relative to the culture from which they derive so that any attempt to formulate postulates that grow out of the beliefs or moral codes of one culture must to that extent detract from the applicability of any Declaration of Human Rights to mankind as a whole.* (AAA 1947, pp. 541–542; italics in the original)

Jarvie argues that these principles are inconsistent with each other and the third principle is inconsistent with itself. The most problematic principle would be the second one, where the respect for cultural differences is justified by the appeal to science as the final court of judgment (Jarvie 2007, p. 573). This corresponds to Herskovits's claim that cultural relativism is a scientific attack on an old philosophical problem (Herskovits 1972, p. 14). Is science "relative to the culture from which they derive"? Or, is it universal? Both Herskovits and the AAA statement are not clear on this point.

After discussing the AAA statement, Jarvie critically assesses cultural relativism. In his opinion, cultural relativism has two major problems: descriptive inaccuracy and logical incoherence. The first problem, descriptive inaccuracy, is that people in most cultures are not cultural relativists, but absolutists. There are many people in the world, but almost all of them would argue that their knowledge of how the world works and their view of morality are absolutely correct, and thus people who differ from them are completely mistaken. For this reason, cultural relativism is descriptively inaccurate. The second problem, logical incoherence, is a well-known and typical criticism of cultural relativism, and thus I need not explain it in detail. It means that cultural relativism cannot be formulated in a logically coherent way (Jarvie 2007, pp. 574–577).

Regardless of these problems, cultural relativism is popular among anthropologists. Jarvie investigates why it is. According to him, when they have to choose between ethnocentrism and cultural relativism, anthropologists select cultural relativism. Since anthropologists think that they must be tolerant of different cultures and values, they cannot accept ethnocentrism. But Jarvie claims that such a dichotomy between ethnocentrism and cultural relativism is a false one, because there is a middle view of self-critical rationality (Ibid., pp. 580–581). In his view, to be tolerant of other cultures, we need not embrace cultural relativism. Jarvie argues as follows: "The critical and open-minded attitude of science is also the best for encouraging the tolerance and respect enjoined by multiculturalism. Not to compare, not to judge, not to debate and criticize the others in the multicultural mosaic is impossible" (Ibid., p. 583). Such a view is not ethnocentric, because it is not accepted dogmatically. But it does not imply cultural relativism because it does not endorse cultural differences as they are; this is why Jarvie regards his view as a middle view.

Jarvie's Social and Institutional View of Rationality

But what does Jarvie's middle view mean, and where does it come from? Considering these questions would help in understanding his social and institutional view of rationality. First, we need to examine Jarvie's argument for weak absolutism. Jarvie distinguishes strong absolutism, weak absolutism, and relativism, and opts for weak absolutism. Jarvie defines these three positions in the following way:

> Strong absolutism is the view that there are only absolute truths. Weak absolutism allows that there are degrees of truth as well as absolute truth, and the former is defined as approximation to the latter.... Relativism is the position that all assessments are assessments relative to some standard or other, and standards derive from cultures. The attempt to assess without regard to cultural context and, particularly, the attempt to assess cognitive statements on some transcendental scale of truth, is futile. No assessment can escape the web of culture and hence all assessment is culturally relative. (Jarvie 1983, pp. 45–46)

Jarvie argues that both strong absolutism and relativism are based on psychologist theories of rationality that regard it as a matter of individual mentality, and they get trapped into dogmatism and intolerance. Strong absolutists are either being confident or skeptical of obtaining truths. As a result, they are not interested in social arrangements that encourage tolerance and criticism to obtain more knowledge. By contrast, relativists want to be tolerant of cultural differences, but they do not take seriously critical assessments of cultures, which intend to promote the growth of knowledge. Thus, they get trapped into dogmatism and intolerance (Ibid., pp. 47–48).

For these reasons, Jarvie cannot accept both strong absolutism and relativism and chooses weak absolutism. In weak absolutism, the growth of knowledge depends not on the minds of individuals but on institutions where they work. By claiming so, Jarvie emphasizes the importance of a social and institutional character of rationality. This can be seen from Jarvie and Agassi's model of workshop rationality, where people exchange ideas and try to solve problems in a collaborative discussion. Some might think that the idea of workshop presupposes a fixed relationship between master and apprentice or teacher and student. That is not the case, however. The master and the apprentice can change their roles in a real discussion, and the former can sometimes accept the latter's proposal. A historical example of workshop would be a Socratic dialogue. Although Plato's dialogues tend to illustrate Socrates as someone who knows the answer beforehand, Jarvie and Agassi argue that if the real Socrates were such a person, he would not have attracted many students. In their view, Socrates attracted students because of his workshop atmosphere (Jarvie and Agassi 1979, p. 360; see also Jarvie 2005, pp. 474–475).

From Jarvie's argument for weak absolutism and his model of workshop rationality, we can discern that Jarvie provides a social and institutional perspective of rationality. This point is elaborated in his view of science as a social institution (Jarvie 2001). What does he mean by a social institution? As far as I know, Jarvie does not seem to define precisely the idea of social institution. This might come from Popper's anti-essentialist view that we should worry not about meanings of words but about problems about facts (Popper 1974, p. 12). True; even if we try to clarify or define words, that could be endless, and thus unfruitful. Still, we need some working definition to study social institutions. In this regard, it would be helpful to look over Popper's usage of the term "social institution." Popper proposes to use it in the following way: social institutions are "all those things which set limits or create obstacles to our movements and actions almost as if they were physical bodies or obstacles. Social institutions are experienced by us as almost literally forming part of the furniture of our habitat" (Popper 1994, p. 167). But such a usage is so broad that many things can be counted as social institutions: language, contract, market, and so on. According to Popper, one important characteristic of social institutions is that most of them are products of unintended consequences of human actions (Popper 1944, p. 122). Jarvie regards this characteristic as important, because he contends that "[s]ocial reality ... is the unintended consequence par excellence" (Jarvie 1998, p. 378, n. 7). Hence, there is no fundamental difference between Popper and Jarvie.

The idea of social institution has been one of the important topics in philosophy and the social sciences. Critical rationalists have examined it for many years (Agassi 1960, 1975; Wisdom 1970; Wettersten 2006). Not only critical rationalists but also other philosophers and social scientists have scrutinized it. Then how is the critical rationalist view of social institutions related to the current philosophical and social scientific discussions about them? In considering this question, Guala's work on institutions would be a good guiding thread. Guala proposes a unified theory of institutions by combining two major theories of institutions: rule-based and equilibrium-based ones. He calls it the rules-in-equilibrium approach and argues that "[i]nstitutions, in a nutshell, are rules that people are motivated to follow" (Guala 2016, p. xxv). According to Guala, the rule-based view sees institutions as rules, and it can be found in works by Max Weber, Talcott Parsons, Friedrich Hayek, Douglas North, and John Searle. By contrast, the equilibrium-based view takes institutions as the equilibrium of strategic games, and its major proponents are Thomas Schelling, David Lewis, and other game theorists. Although these views are sometimes regarded as incompatible, Guala argues that they are complimentary.

How would Jarvie respond to Guala's complimentary view? In some of his works, Jarvie suggests that he regards Popper's view as one of the equilibrium-based views. For instance, he claims that "[f]or Popper, institutions are intermediate between persons: they serve as means of co-ordination between the diverse aims of diverse people" (Jarvie 1972, p. 32). Elsewhere, while criticizing Searle's disregard for unintended consequences, Jarvie contends that Agassi anticipates Lewis's view of institutions as coordinating mechanisms (Jarvie 1998, pp. 378–379, n. 7). From these remarks, it is not clear what Jarvie would think about Guala's view. But by comparing their views, we can discern some differences between them.

One of the differences is the status of the unintended consequence. Jarvie stresses the importance of the unintended consequence, while Guala scarcely discusses it. Another difference is their views of functionalism. Guala regards institutions as solutions to coordination problems that people face (Guala 2016, p. 185). This is welcoming to critical rationalists, because they accept Popper's emphasis on problem-solving. But at the same time, Guala claims that "the equilibrium model will support a functional explanation of the form: 'the rules exist because they help people solve a coordination problem'" (Ibid., p. 30). Hence, he provides a functional explanation of institutions. By contrast, Jarvie has an ambivalent attitude toward functionalism, because he contends that functionalism is false (or tautologically true) as a theory, but it is fruitful as a method. He states that "a theory tells us the way the world is (all societies are cohesive); a method tells us how to explore the world (look at the context of institutions)" (Jarvie 1973, p. 35). Elsewhere he values functionalist questions focusing on institutions, but does not value functionalist explanations (Jarvie 2003, p. 268). From these claims, we can infer that Jarvie would not accept Guala's functional explanation of institutions. Although I cannot pursue this problem further, how Jarvie's view is related to the recent philosophical and social scientific investigations into social institutions deserves a serious examination.

Conclusion

In this chapter, I have examined Jarvie's view of rationality and his criticism of cultural relativism. He argues that rationality can be seen as a matter of degree. By doing so, he suggests a middle way to avoid pitfalls of ethnocentrism and cultural relativism. This is an important progress. But it becomes clear that Jarvie's idea of social institution needs to be explicated further, and the relation between his view and the recent philosophical and social scientific views of social institutions awaits further examination. These are problems not only for Jarvie but also for people who value his contributions to the philosophy of the social sciences. These problems need to be discussed and examined in workshops where tolerance, respect, and criticism are encouraged.

References

Agassi, Joseph. 1960. Methodological Individualism. *British Journal of Sociology* 11 (3): 244–270.

———. 1975. Institutional Individualism. *British Journal of Sociology* 26 (2): 144–155.

American Anthropological Association (AAA). 1947. Statement on Human Rights. *American Anthropologist* 49 (4): 539–543.

Bartley, William W. 1968. Theories of Demarcation Between Science and Metaphysics. In *Problems in the Philosophy of Science*, ed. Imre Lakatos and Alan Musgrave, 40–64; reply: 102–119. Amsterdam: North-Holland.

———. 1984 [1962]. *The Retreat to Commitment*. La Salle, IL: Open Court.

Bidney, David. 1953. The Concept of Value in Modern Anthropology. In *Anthropology Today: An Encyclopedic Inventory*, ed. Alfred L. Kroeber, 682–699. Chicago: University of Chicago Press.

Descartes, René. 1911 [1637]. Discourse on the Method of Rightly Conducting the Reason and Seeking for Truth in the Sciences. In *The Philosophical Works of Descartes*. Translated by Elizabeth S. Haldane and George R. T. Ross, vol. 1, 79–130. London: Cambridge University Press.

Engle, Karen. 2001. From Skepticism to Embrace: Human Rights and the American Anthropological Association from 1947–1999. *Human Rights Quarterly* 23 (3): 536–559.

Guala, Francesco. 2016. *Understanding Institutions: The Science and Philosophy of Living Together*. Princeton, NJ: Princeton University Press.

Herskovits, Melville J. 1972. *Cultural Relativism: Perspectives in Cultural Pluralism*. New York: Random House.

Jarvie, Ian C. 1970. Understanding and Explanation in Sociology and Social Anthropology. In *Explanation in the Behavioural Sciences*, ed. Robert Borger and Frank Cioffi, 231–248; reply: 260–269. London: Cambridge University Press.

———. 1972. *Concepts and Society*. London: Routledge & Kegan Paul.

———. 1973. *Functionalism*. Minneapolis, MN: Burgess.

———. 1983. Rationality and Relativism. *British Journal of Sociology* 34 (1): 44–60.

———. 1984. *Rationality and Relativism: In Search of a Philosophy and History of Anthropology*. London: Routledge & Kegan Paul.

———. 1998. Situational Logic and Its Reception. *Philosophy of the Social Sciences* 28 (3): 365–380.

———. 2001. *The Republic of Science: The Emergence of Popper's Social View of Science, 1935–1945*. Amsterdam: Rodopi.

———. 2003. Fuller on Science. *Philosophy of the Social Sciences* 33 (2): 261–285.

———. 2005. Workshop Rationality, Dogmatism, and Models of the Mind. In *The Mind as a Scientific Object: Between Brain and Culture*, ed. Christina E. Erneling and David M. Johnson, 471–486. Oxford: Oxford University Press.

———. 2007. Relativism and Historicism. In *Philosophy of Anthropology and Sociology*, ed. Stephen P. Turner and Mark W. Risjord, 553–589. Amsterdam: Elsevier.

Jarvie, Ian C., and Joseph Agassi. 1979. The Rationality of Dogmatism. In *Rationality Today*, ed. Theodore F. Geraets, 353–362. Ottawa: University of Ottawa Press.

Kuhn, Thomas S. 1996 [1962]. *The Structure of Scientific Revolutions*. Chicago: University of Chicago Press.

Popper, Karl R. 1940. What Is Dialectic? *Mind* 49: 403–426.

———. 1944. The Poverty of Historicism, II: A Criticism of Historicist Methods. *Economica* 11 (43): 119–137.

———. 1945. *The Open Society and Its Enemies*. 2 vols. London: George Routledge and Sons.

———. 1949. Towards a Rational Theory of Tradition. In *The Rationalist Annual 1949*, ed. Frederick Watts, 36–55. London: Watts and Co.

———. 1974. Autobiography of Karl Popper. In *The Philosophy of Karl Popper*, ed. Paul A. Schilpp, vol. 1, 1–181. La Salle, IL: Open Court.

———. 1994. Models, Instruments, and Truth: The Status of the Rationality Principle in the Social Sciences. In *The Myth of the Framework: In Defence of Science and Rationality*, ed. Mark A. Notturno, 154–184. London: Routledge.

Shearmur, Jeremy. 2017. Popper's Influence on the Social Sciences. In *The Routledge Companion to Philosophy of Social Science*, ed. Lee McIntyre and Alex Rosenberg, 55–64. London: Routledge.

Wettersten, John. 2006. *How Do Institutions Steer Events? An Inquiry into the Limits and Possibilities of Rational Thought and Action*. Aldershot: Ashgate.

Winch, Peter. 1958. *The Idea of a Social Science and Its Relation to Philosophy*. London: Routledge and Kegan Paul.

———. 1964. Understanding a Primitive Society. *American Philosophical Quarterly* 1 (4): 307–324.

———. 1970. Comment [on Jarvie 1970]. In *Explanation in the Behavioural Sciences*, ed. Robert Borger and Frank Cioffi, 249–259. London: Cambridge University Press.

Wisdom, John O. 1970. Situational Individualism and the Emergent Group-Properties. In *Explanation in the Behavioural Sciences*, ed. Robert Borger and Frank Cioffi, 271–296; reply: 306–311. London: Cambridge University Press.

Yoshida, Kei. 2014. *Rationality and Cultural Interpretivism: A Critical Assessment of Failed Solutions*. Lanham, MD: Lexington Books.

PART III

The Popperian Legacy

CHAPTER 14

Popper's Institutional Turn

Rafe Champion

Introduction

For English-speaking people, Popper first signalled what Jarvie called his "social turn" in *The Open Society and Its Enemies* (1966, Chap. 23) and in *The Poverty of Historicism* (1963a, Sects. 31–32). Popper confronted Karl Mannheim's exposition of the Marxist doctrine that our beliefs are determined by class interest and by the social and historical situation of our time. In defence of scientific objectivity, Popper turned the sociology of knowledge on its head by arguing that its focus on the origin of subjective beliefs did not engage with its proper object of inquiry, namely, knowledge as a public or social product. He claimed that the objectivity of science comes from the process of free criticism in the scientific community: "It may be said that what we call 'scientific objectivity' is not a product of the individual scientist's impartiality, but a product of the social or public character of scientific method; and the individual scientist's impartiality is, so far as it exists, not the source but rather the result of this socially or institutionally organized objectivity of science" (Popper 1966, p. 220). Hence scientific objectivity is a situational or institutional problem which calls for such things as theoretical pluralism, clear formulation of the problems that the theories are supposed to solve, the design of critical

R. Champion (✉)
Independent Scholar, Sydney, NSW, Australia

R. Sassower, N. Laor (eds.), *The Impact of Critical Rationalism*,
https://doi.org/10.1007/978-3-319-90826-7_14

experiments, and the existence of journals, seminars, and conferences to facilitate critical discussion. Some of these requirements are provided by individual scientists, especially new ideas and imaginative criticism, while others call for institutions, including political institutions to maintain the autonomy of the journals and the research institutes.

In *The Poverty of Historicism* Popper carried the institutional analysis further in his argument against the psychological approach of Auguste Comte and John S. Mill to explain scientific and industrial progress. The institutional approach points to the work of Douglas North who states: "Institutions form the incentive structure of a society and the political and economic institutions, in consequence, are the underlying determinants of economic performance" (North 1993). Popper adopted the counter-intuitive procedure of trying to imagine conditions under which progress would be arrested.

> By closing down or controlling laboratories for research, by suppressing or controlling scientific periodicals and other means of discussion, by suppressing scientific congresses and conferences, by suppressing Universities and other schools, by suppressing books, the printing press, writing, and, in the end, speaking. All these things which indeed might be suppressed (or controlled) are social institutions .... Scientific method itself has social aspects. Science, and more especially scientific progress, are the results not of isolated efforts but of the free competition of thought. For science needs ever more competition between hypotheses and ever more rigorous tests. And the competing hypotheses need personal representation, as it were: they need advocates, they need a jury, and even a public. This personal representation must be institutionally organized if we wish to ensure that it works. (Popper 1963a, pp. 154–155)

Jarvie's *The Republic of Science*

Jarvie's *The Republic of Science* aroused little interest when it was published in 2001, too late for mention in Steve Fuller's *The Governance of Science* (2000). John Wettersten wrote a critical appraisal (2006), and Hugh Heclo acknowledged it as an influence in *On Thinking Institutionally* (2008). Jarvie's dust jacket announces that "This book offers a careful re-reading of Popper's classic falsificationist demarcation of science, stressing its institutional aspects. Ian Jarvie tracks Popper's social thinking about science, individuals, institutions, and rationality through *The Poverty of Historicism* and *The Open Society and Its Enemies* as he criticised and improved his earlier work. New links are established between the works of the 1935–1945 period, revealing them as a source for criticism of the institutions and governance of science." Jarvie found a "social turn" in Popper's first published work. He clearly took to heart one of Popper's favourite catch phrases "We never know what we are saying": our theories have contents and consequences that we do not realize. He suggested that Popper, for all his resistance to the sociological approach, in fact anticipated it. Some comment is required regarding Jarvie's use of the term "social turn," because it has created a great deal of confusion: is there a change in Popper's thinking or a deviation from the logical positivists? It may be more helpful to talk about the *social or institutional theme* in Popper's work, and so my gloss on Jarvie's thesis is that a significant achievement of *Logik der Forschung* was to emphasize the need for a critical approach to methodological procedures and conventions which function as rules of the game in science. That aspect of the book was generally overlooked because commentators and critics concentrated on the concerns which Popper shared with his opponents, the problems of demarcation and induction.

For those who are equally interested in Popper's work in the philosophy of science and politics, the critical approach to the rules of the game is common to both. One of the functions of the philosophy of science is to formulate and criticize the rules of the game of science, while political philosophers may do the same for the rules of social and political life. Some of these rules are written laws, regulations, protocols, and procedures, and others are unwritten customs and habitual practices. In science, these would appear to include what Michael Polanyi called "tacit knowledge." Jarvie's Introduction, "Science as an Institution," sets out the major issues in the complex relationship between science and society. The word science may refer to a body of public knowledge, a set of beliefs about the world, the whole range of activities performed by scientists, and the complex of social and political institutions which support and influence the activities of scientists. Jarvie offers a survey of various approaches to science, including the positivist and falsificationist demarcation principles, and Merton's account of the distinguishing features of scientific knowledge.

Jarvie unpacks the hidden elements of the "social turn" in Popper's early philosophy of science. He identifies a number of procedural rules which constitute what he calls Popper's "proto-constitution of science." This is the foundation of his project, and it is spelled out in some detail: "My argument will be that thinking socially (rather than logically or psychologically) is central to Popper's philosophical enterprise beginning with *The Logic of Scientific Discovery*, continuing for his ten most creative years, and emerging sporadically after that. Popper's consistent ability to think socially also does much to account for his originality, since it is hard to do and its difficulty is attested by how often readers and critics of Popper do not grasp that this is what he is doing" (2001, p. 21). Jarvie argues that the roots of Popper's social/institutional thinking can be traced back to *Logik der Forschung*, but after *The Open Society and Its Enemies*, Popper did not expand on the theme. Wettersten suggests that Popper "was so concerned not to explain away scientific knowledge as a mere social phenomenon that he did not engage in the social studies of science even though his view called for such studies" (Wettersten 2006). This looks like an example of "snow blindness," a term which Arthur Koestler uses to refer to "that remarkable form of blindness which often prevents the original thinker from perceiving the meaning and significance of *his own* discovery" (Koestler 1967, pp. 217–221). Jarvie continues to outline "Popper's 1935 Proto-Constitution for the Republic of Science." This contains Popper's supreme or meta-rule (SR) and 14 subsidiary rules which constitute the rudimentary scaffolding for Popper's republic of science. Here they are:

> SR: The other rules of scientific procedure must be designed in such a way that they do not protect any statement in science against falsification.
>
> R1: The game of science is, in principle, without end. He who decides one day that scientific statements do not call for any further test, and that they can be regarded as finally verified, retires from the game.
>
> R2. Once a hypothesis has been proposed and tested, and has proved its mettle, it may not be allowed to drop out without "good reason." (Popper 1972, pp. 53–54)

The "supreme rule" and the first two subsidiary rules were proposed by Popper, and Jarvie identified additional rules in the text of Popper's book:

> R3. We are not to abandon the search for universal laws and for a coherent theoretical system, nor ever give up our attempts to explain causally any kind of event we can describe. (Ibid., p. 61)
>
> R4. I shall … adopt a rule not to use undefined concepts as if they were implicitly defined. (Ibid., p. 75)
>
> R5. Only those auxiliary hypotheses are acceptable whose introduction does not diminish the degree of falsifiability or testability of the system in question but, on the contrary, increases it. (Ibid., p. 83)

> R6. We shall forbid surreptitious alterations of usage. (Ibid., p. 84)
>
> R7. Inter-subjectively testable experiments are either to be accepted, or to be rejected in the light of counter-experiments. (Ibid.)
>
> R8. The bare appeal to logical derivations to be discovered in future can be disregarded. (Ibid.)
>
> R9. After having produced some criticism of a rival theory, we should always make a serious attempt to apply this criticism to our own theory. (Ibid., 85n)
>
> R10. We should not accept stray basic statements, i.e., logically disconnected ones—but ... we should accept basic statements in the course of testing theories; or raising searching questions about these theories, to be answered by the acceptance of basic statements. (Ibid., p. 106)
>
> R11. This makes our methodological rule that those theories should be given preference which can be most severely tested ... equivalent to a rule favouring theories with the highest possible empirical content. (Ibid., p. 121)
>
> R12. I propose that we take the methodological decision never to explain physical effects, i.e. reproducible regularities, as accumulations of accidents. (Ibid., p. 199)
>
> R13. A rule ... which might demand that the agreement between basic statements and the probability estimate should conform to some minimum standard. Thus, the rule might draw some arbitrary line and decree that only reasonably representative segments (or reasonably "fair samples") are "permitted," while a-typical or non-representative segments are "forbidden." (Ibid., p. 204)
>
> R14. The rule that we should see whether we can simplify or generalize or unify our theories by employing explanatory hypotheses of the type mentioned (that is to say, hypotheses explaining observable effects as summations or integrations of micro events). (Ibid., p. 207)

Jarvie's commentary on the "constitution" begins with the reminder that it is very incomplete and very abstract, as though the whole of science is a kind of debating society, leaving out of account a great deal of gritty reality, such as the question of leadership. He noted Polanyi's discussion in *Personal Knowledge* (1958) of rules in scientific practice which he called maxims. Polanyi considers them "tacit," as they are picked up in the process of training and induction into the scientific community. They are interpreted and administered by the leadership of the scientific community, but they do not include any guidelines for critical appraisal and reform of the maxims even in the republic itself, much less by outsiders. Jarvie continues to take up three topics in *The Poverty of Historicism* (1963a): first,

Popper's ideas about the emergence of institutions, second, his ideas about individuals and their function in science, and third, his ideas on rationality and the scope for objectivity and testing in the human sciences. Popper was influenced by the ideas of Carl Menger and the Austrian School of social and economic thought which were "in the air" in Vienna at the time. Two of the central Austrian doctrines which Popper took on board are the theory of the origin of social institutions and methodological individualism.

Jarvie also aims to capture Popper's published ideas over the decade 1935–1945, where a problem of organization becomes apparent, partly due to the density of argument. Popper spent several hundred pages to spell out his arguments and Jarvie in a hundred pages set out to convey the gist of those arguments along with his first take on the way that Popper's ideas about society enrich his ideas about science and vice versa. Jarvie offers the thesis on Popper's "social turn," the analogy between Popper's approach to the rules of the game in science and society, the parallels between Popper and the Austrians in the methodology of the social sciences, and the implications of all of the above in working out the relationships between science (in all its various aspects) and society (and politics). Many issues arise from the troubled relationship between science and politics since applied science became vital for national defence and pure science moved beyond the point where people could win Nobel Prizes with equipment purchased from the local hardware shop. Much of Jarvie's discussion on these matters is directly or indirectly concerned with the governance of science.

The Social/Institutional Approach by Popper

There are three implications and applications of the social/institutional approach by Popper and others: first, a common thread in Popper's philosophy of science and his approach to social and political issues, second, a common thread in Popper and Hayek, and third, the governance of science. Regarding the first and second, Popper's criterion for the value of his proposals in *Logik der Forschung* was their utility in helping scientists to get on with their work. He didn't quite put it like that, he referred to "their fertility–their power to elucidate the problems of the theory of knowledge" (1972, p. 38). The point is that working scientists are disadvantaged if they operate with dysfunctional ideas in the theory of knowledge, such the notion that they should start with observations. The point is to begin with the right questions, questions which facilitate purposeful and effective action both in the experimental sciences and in social and political matters. Collingwood made the same point in writing about "the logic of question and answer" (1957, Chap. V). Working scientists are expected to carry out experiments and make observations. Popper's proposals (rules) can be applied to their activities, including their deliberations before and after experimental and observational work. In the realm of politics and social action, Popper similarly provided some rules, starting with the suggestion to use the language of political demands or proposals instead of the language of essentialism and historicism.

While the essentialist wants to find the essence of justice (or the state) and the historicist wants to look at the history of the state or its operations, Popper suggested that we should consider *what we want* from our system of justice and from the state:

> In a clear presentation of this theory [of the protective state], *the language of political demands or of political proposals should be used*; that is to say, we should not try to answer the essentialist question: What is the state, what is its true nature, its real meaning? Nor should we try to answer the historicist question: How did the state originate, and what is the origin of political obligation? We should rather put our question in this way: *What do we demand from a state? What do we propose to consider as the legitimate aim of state activity*? (Popper 1966, p. 109; italics added)

That is not to dismiss studies which search for patterns or regularities in social or political processes. Similarly, we do not dismiss historical studies. Instead, this means that we do not confuse trends with laws, and we do not have to submit to historical tendencies, like the ruin of democracy under communism in Russia or state socialism in Germany.

Hayek took a similar approach in *The Constitution of Liberty* (1976) where he discussed the pros and cons of democracy. He noted that the power of the state was not expected to be a problem after power came into the hands of the people, because the new rulers (replacing kings, emperors, and despots) would not harm them. However, the people are represented by politicians, parties, and factions which are not under the immediate control of the people. Hence the danger of the abuse of power persists under democracy: "But it is not democracy but unlimited government which is objectionable .... *It is not who governs but what government is entitled to do that seems to me the essential problem*" (p. 403; italics added).

Compare this with Popper's criticism of the idea that the question of sovereignty (who shall rule) is a fundamental issue in politics: "It is high time for us to learn that the question '*who* is to wield the power in the state?' matters little as compared with the questions '*how* is the power wielded' and '*how much* power is wielded?'" (Popper 1966, p. 162). Writing on Marxism and revolutionary violence, Popper pointed to the need for "rules of engagement" for the use of violence by the state both in defence of the realm and in policing law and order. This approach calls for rules for revolutionary violence in the extreme situation where the rulers cannot be dismissed by democratic means.

Hayek's last book *The Fatal Conceit* (1991) pulled together the threads of his project on rationality, law, and liberty. He argued for the importance of evolved moral conventions which he called the "extended moral order" of Western civilization, including rules related to markets, private property, honesty, contracts, and privacy. Popper also referred to the vital function of the moral framework of society (Popper 1963b, Chap. 17), and more recently Deirdre McCloskey has written extensively on the "the bourgeois virtues" and related matters (McCloskey 2006, 2010). Popper and Hayek improved on those principles which function as the rules of the game in social life. As I have argued before:

> The 'rules of the game' range from the possibly innate rules of grammar, through the tacit knowledge of local traditions and folkways to the rules of games and other codified forms of procedure. They include the laws of the land embodied in common law, statutes and constitutions. At another level they include the unformalized rules of the moral framework which Popper noted in his essay "Public Opinion and Liberal Principles". The study of these rules would need to probe the way that different sets of rules support or undermine each other and the effect of changing from one set to another. This would be essentially an ecological study with the emphasis on unintended 'downstream' effects of changes in the prevailing order. This does not imply a rigidly conservative attitude to the status quo, it merely signals that we need to learn from our conscious or unconscious social experiments. This approach would supplement the methods of conceptual analysis and crude 'positivist' empirical description of social and political systems. It would have the theoretical advantage of linking disciplines and the practical merit of being continually in touch with problems and their possible solutions. (Champion 1992)

The Governance of Science

Popper's reference to the institutional context of science in *The Poverty of Historicism* points to the governance of science which became a big issue in Britain when Marxists, such as John Bernal (1939), pushed for government control of the national research effort. This provoked a reaction to defend autonomous science led by Polanyi (1962). For an account of this episode, see Jerome Ravetz (1971). Other postwar developments intensified the issue of governance, notably the rise of Big sScience and the official discovery of the phenomenon of "normal science." Big Science refers to the major research and development projects on the model of the Manhattan Project which transformed the scientific and academic landscape (Shills 1972; Greenberg 2001). In his unpublished lectures, Popper referred to the menace of Big Science with too much money chasing too few ideas and other adverse effects. Speaking from a very different vantage point, President Eisenhower in his farewell speech to the nation in 1961 warned that the nation's scholars might be dominated by Federal employment and project allocations: "the power of money is ever present–and is gravely to be regarded. Yet, in holding scientific research and discovery in respect, as we should, we must also be alert to the equal and opposite danger that public policy could itself become the captive of a scientific-technological elite." Then came Thomas Kuhn's account of normal science which was taken up in some quarters as a recommendation about the proper way to go, especially in the social sciences which were keen to copy the more prestigious disciplines (Notturno 1984).

A more recent development is a rapidly growing literature on problems in the quality of published research. Richard Horton in his capacity as editor-in-chief of *The Lancet* wrote: "The case against science is straightforward: much of the scientific literature, perhaps half, may be simply untrue .... Science has taken a turn towards darkness" with reference to small sample sizes, invalid analyses, conflicts of interest, and obsession with fashionable trends (Horton 2015). There is concern about the increasing incidents of retractions and the higher rate of retractions in high-impact journals (Fang et al. 2011) and the dangerous liaison of science and politics (Butos and McQuade 2006). Less than 12% of articles in 2004 in the *Journal of Economic Theory* passed three tests: stating a theory, explaining why it mattered, and testing it (Klein and Romero 2007). There are problems of replication of results and politicization in some fields, such as climate science. Another concern is the declining publication of negative results (Fanelli 2012). It would be interesting to explore if this has any basis in the persistent teaching of confirmation theory rather than critical rationalism in epistemology and the philosophy of science.

The economist Gordon Tullock was inspired by his contact with Popper to make a significant but little appreciated contribution to the literature on the governance of science. In *The Organization of Inquiry* (1966), he approaches the topic of scientific research and publication as a student of legal, social, and economic systems to sketch a scenario for the decline of a scientific discipline, given a particular combination of motivational factors and institutional incentives. Concerning the people who engage in pure and applied research, he identified three kinds of curiosity: first, pure curiosity and compulsion to find how the world works, which drives the great scientists; second, the passionate desire to solve practical problems, which drives the great builders and engineers; and third, "induced curiosity" directed to either pure or applied problems. The researchers with induced curiosity are those who do not have a consuming passion for research but do it as a job. Outstanding work can be produced by academics seeking promotion and even by nine to five scientists, but Tullock's analysis addresses tendencies which could emerge in a system where more and more of the workers have "induced" curiosity and less pure curiosity.

Closely related to the motives of the investigators are their concern for the quality of the work and their willingness to test their assumptions and their results. Tullock notes that dedicated truth seekers and serious problem-solvers must pay close attention to reality to align their ideas with it, and this demands constant testing and critical evaluation. In contrast, researchers only aiming to publish to satisfy requirements can be happy with results that are merely publishable even if they are not robust. As Tullock puts it, scientific concern with the real world can run second to other matters: "If he could establish and maintain his reputation, and hence his job, by reporting completely fictional discoveries, this would accomplish his end. While an investigator motivated by curiosity or practical utility must, of necessity, concern himself with real phenomena, the man motivated by induced curiosity could, if the risk of discovery were not great, simply ignore reality" (1965, p. 56). Though the peer-review process is designed to improve matters, this may not happen when reviewers are too closely associated with the authors. Tullock speculates that this would be most likely to happen in a field dominated by "induced" and "normal scientists" rather than the dedicated truth seekers. In his words:

"The concern with reality that unites the sciences, then, may be absent in this area, and the whole thing may be reduced to a pseudo-science like genetics in Lysenko's Russia ... these symptoms may be found in some of the social sciences" (Ibid., p. 56).

Conclusion

We can be grateful for the contribution of Jarvie and Tullock for their wide-ranging contributions to their respective fields and for the two books cited for special mention here. It seems that Jarvie's book did not arouse the interest which I think is warranted, especially regarding the implications and applications of the social/institutional approach to research. Let us ensure that there is a more expansive and inclusive conversation on the themes and issues which Jarvie and Tullock addressed, especially in relation to the governance of science. This could generate more heat than light, but it will be an opportunity lost if it does not occur.

References

Bernal, John D. 1939. *The Social Function of Science*. London: Routledge.

Butos, William N., and Thomas J. McQuade. 2006. Government and Science: A Dangerous Liaison? *The Independent Review* 11 (2): 177–208.

Champion, Rafe A. 1992. Review of *the Fatal Conceit. Melbourne Age Monthly Review*, November: 17–19.

Collingwood, Robin G. 1957. *An Autobiography*. Oxford: Oxford University Press.

Fanelli, Dominic. 2012. Negative Results Are Disappearing from Most Disciplines and Countries. *Sociometrics* 90 (3): 891–904.

Fang, Ferric, Arturo Casadevall, and Richard P. Morrison 2011. Retracted Science and the Retraction Index. *Infectious Immunology* 78 (10): 3855–3859. http://iai.asm.org/content/79/10/3855.

Fuller, Steve. 2000. *The Governance of Science*. Buckingham: Open University Press.

Greenberg, David S. 2001. *Science, Money, and Politics: Political Triumph and Ethical Erosion*. Chicago: Chicago University Press.

Hayek, Friedrich A. 1976. *The Constitution of Liberty*. London: Routledge.

———. 1991. *The Fatal Conceit*. Chicago: University of Chicago Press.

Heclo, Hugh. 2008. *On Thinking Institutionally*. London: Paradigm Publishing.

Horton, Richard. 2015. Offline: What Is Medicine's 5 Sigma? *The Lancet* vol. 385.

Jarvie, Ian C. 2001. *The Republic of Science: The Emergence of Popper's Social View of Science 1935–1945.* Amsterdam: Ripodi.

Klein, Daniel B., and Pedro Romero. 2007. Model Building Versus Theorising: The Paucity of Theory. *The Journal of Economic Theory* 4 (2): 241–271.

Koestler, Arthur. 1967. *The Act of Creation.* London: Pan.

McCloskey, Diedre N. 2006. *The Bourgeois Virtues: Ethics for an Age of Commerce.* Chicago: University of Chicago Press.

———. 2010. *Bourgeois Dignity: Why Economics Can't Explain the Modern World.* Chicago: University of Chicago Press.

North, Douglass C. 1993. Economic Performance Through Time. http://www.nobelprize.org/nobel_prizes/economic-sciences/laureates/1993/north-lecture.html.

Notturno, Mark. 1984. The Popper/Kuhn Debate: Truth and the Two Faces of Relativism. *Psychological Medicine* 14: 273–289.

Polanyi, Michael. 1962. The Republic of Science: Its Political and Economic Theory. *Minerva* 1: 54–73.

———. 1958. *Personal Knowledge.* London: Routledge.

Popper, Karl R. 1966. *The Open Society and Its Enemies.* London: Routledge.

———. 1963a. *The Poverty of Historicism.* London: Routledge.

———. 1963b. *Conjectures and Refutations: The Growth of Scientific Knowledge.* London: Routledge.

———. 1972. *The Logic of Scientific Discovery.* London: Hutchinson.

Ravetz, Jerold. 1971. *Scientific Knowledge and Its Social Problems.* Harmondsworth: Penguin Books.

Shills, Edward. 1972. *The Intellectuals and the Powers, and Other Essays.* Chicago: University of Chicago Press.

Tullock, Gordon. 1965. *The Organization of Inquiry.* Durham, NC: Duke University Press.

Wettersten, John. 2006. Essay Review of *The Republic of Science. Philosophy of Science* 73 (1): 108–121.

CHAPTER 15

The Republic of Science and Its Citizens: What Role May Humanities Play Within the Popperian Framework?

José A. Colen and Scott Nelson

INTRODUCTION

Jarvie is a versatile and prolific sociologist whose works are both provocative and sharp. He has written an incredible number of important papers and books in various fields. Since no one can follow him in all these fields, one might be driven to despair. This chapter cannot therefore but aim at a very limited goal, merely underlining his specific contribution to the study of some of Popper's ideas in *The Poverty of Historicism* (1957) and showing how he goes further than Popper. We will thus seek to address a concrete issue: the unity of the method of the natural and social sciences. As Jarvie acknowledges, "the difficulty of writing about Popper's view of the

J. A. Colen (✉)
Fundação para a Ciência e a Tecnologia, Centre for Ethics, Politics and Society, Minho University, Braga, Portugal

ICS – University of Navarra, Navarra, Spain
e-mail: jcolen@unav.es

S. Nelson
Independent Scholar, Vienna, Austria

R. Sassower, N. Laor (eds.), *The Impact of Critical Rationalism*,
https://doi.org/10.1007/978-3-319-90826-7_15

differences between the natural and the social sciences ... has partly to do with the structure of Popper's main work on them, *The Poverty of Historicism*" (1986, p. 71).

Looking through Hayek's subjectivism, we will claim that the difference in method is greater than it seems and that something akin to a Republic of Letters, taking its cue from the historical sciences and humanities, should be built alongside the "Republic of Science," without ignoring the inherent faults in these two republics imposed haphazardly on society at large that Jarvie underlines in a recent book.

The Arguments in Favor of Anti-naturalism in *The Poverty of Historicism*

Jarvie emphasizes that when *The Poverty of Historicism* came out in book form during the fall of 1957, "it caused ripples," and since then "it has become an anticlassic: read but not praised" (Ibid., p. 70). The classics are, as Mark Twain notes in *South of the Equator*, the books everyone talks about but no one really reads. Although the book parallels *Logik*, it is awkward in the sense that the "book begins not with a problem but with a doctrine: historicism, that is, that there are historical laws of social development. The book is a relentless critique of all versions of historicism. Popper suggests that this doctrine is widely diffused in the intellectual atmosphere of our time-acknowledged and, more often than not, unacknowledged—and that it underpins totalitarianism. His critique is to the effect that it is a philosophy of the social replete with error, advocating a method of tackling social problems that will make things worse, not better" (Ibid., p. 71).

In the same text quoted above, Jarvie's purpose is to "restructure *The Poverty of Historicism* on the same lines as *Logik der Forschung*," thereby suggesting that the main problems to be addressed in *The Poverty of Historicism* are not Popper's criticisms of historicism as such, but rather the answers he outlines in response to the following questions: "what are the methods of the social sciences? How, if at all, do the methods of the social sciences differ from the methods of the natural sciences?" (Ibid., p. 73). In accordance with Popper's habit of giving his intellectual opponents the best arguments possible (worthy of attack), *The Poverty of Historicism* begins by enumerating arguments in favor of anti-naturalism. These "negative" doctrines affirm that there are profound differences between sociology and physics relating to the possibility of generalization and the use of the experimental method. To these, some add the complexity of social phenomena, difficulties of accurate forecasting, and the importance of essences (Popper 1957, pp. 5–6). Other arguments emerge throughout the Popperian text, which developed over a dozen years (Brudny 2002, p. 105; Hacohen 2002, p. 352). Popper does not deny that there are "clear" differences, both between the different natural sciences and within the human sciences (e.g., linguistics and economics), but he argues that "the methods in the two fields are fundamentally the same" (Popper 1957, p. 131): the deductive causal method and experience by means of prediction. Even with certain limitations this method can be applied to the historical sciences (Ibid., p. 143).

Among the many arguments that Popper would term anti-naturalist or negative, we would outline these in particular: (a) There are no social laws of universal validity. The possibility of generalization or induction in physics presupposes the uniformity of nature and is inapplicable to the social sciences. In his words: "A method which ignores this limitation" presupposing sempiternal regularities produces a "false and dangerously misleading sociological theory" (Ibid., pp. 6–7). This theory conceals the "intention" of "fostering of a general feeling of inevitability," in particular the futility of the intervention of the law vis-à-vis the "'inexorable laws' of economics" (Ibid., p. 7), when social uniformities are created by man and it is human strength that alters them.

(b) One cannot use the experimental method in sociology. Isolation is not possible in the human sciences, without losing what is more important, and social experiences "can never be repeated under the same precise conditions." Isolation in physics is possible in experiments under the assumption that "similar things will happen" (Ibid., p. 8). He continues: "The sum of the conditions–internal and external–under which we repeat a certain experience" (Ibid., pp. 23–24) may make us "unable to formulate *general laws* which would serve as a description" (Ibid., p. 11).

(c) The complexity of social phenomena prevents one from discovering its constants. We feel that there is something human behind the variations, but the conclusions are more negative than positive. In his words: "Even if there were immutable sociological uniformities, like the uniformities in the field of physics, we might very well be unable to find them" (Ibid., p. 12).

(d) An exact prediction is contradictory, and the interference of the observer's values prevents objectivity in the human sciences. First, because information influences the situations. This fact is labeled by Popper with humor, considering the pseudoscientific Freudian complex the "Oedipus effect": the "interconnection between predictions and the predicted events" means that "an exact and detailed calendar of social events is self-contradictory" (Ibid., pp. 13–14). Secondly, because the subject and object belong to the same world, an observer is, even if involuntarily, influenced by his preferences, and there are as many points of view as there are interests (Ibid., p. 16).

(e) There are no sociological formulas articulated in quantitative terms. It is true that the social sciences successfully use quantitative methods but not in the same way that physics uses quantitative terms and formulas; nothing in the human realm "can be compared to the mathematically formulated causal laws of physics" (Ibid., Chap. 9, p. 25).

(f) In the social sciences, the terms have meanings and correspond to "essences."

Finally, (g) another reason that makes the procedures of the social sciences so different is that they are concerned with the "essential identity of the institution" (Ibid., Chap. 10, pp. 31–32), that is, in social phenomena there can be no mere nominalist description. Nominalism here means describing "how things behave"; words are "merely *useful instruments of description*" (Ibid., p. 29). The social sciences do not use labels but are interested in the intrinsic, intelligible properties of human life (state, nation, class), because the actors of history themselves ascribe or discover "meaning."

Popper has a large arsenal of arguments *against* the possibility of extending the methods of the natural sciences to history and the human sciences (Colen 2010). The list is even more extensive than the one presented in Hayek's brilliant *The Counter-Revolution of Science* (Hayek 1964). Unlike Hayek, however, Popper presents these arguments in order to better *refute* them and does not oppose the transposition of the (basic) methods of physics into sociology, nor does he fear that such practice is at the root of the manifold evils afflicting social theory. The presentation of the anti-naturalistic theses (or fallacies) is at first glance deliberately sympathetic. Moreover, Popper himself constructs the arguments in a seemingly logical way and at various times seems to accept that there is some truth underlying these arguments. He acknowledges, for instance, that the thesis that there are immutable human laws is not always disinterested. While he argues that science is essentially disinterested, he agrees that many scientific advances do not come from "idle curiosity," (Ibid., p. 55) or from a "futile subtlety" (Ibid., p. 57), but from attempts to solve concrete questions, and he agrees to place himself in the practical realm of historicistic "activism." As he asserts, many of the social or public problems have "a practical urgency surpassing even that of cancer research" (Ibid., p. 56).

Anti-naturalism Refuted?

Popper postpones the critique of the anti-naturalistic doctrines to the third part of his book. The exceptions are a lengthy critique of methodological essentialism and the apology of a certain nominalism, the scope and meaning of which, nonetheless, he later clarifies (Shearmur 1996, p. 124; Popper 2006, Chaps. 6 and 7). The reader may guess that a powerful refutation of the arguments will follow. This never happens, as Jarvie notes. Popper's criticisms of the anti-naturalist arguments are surprisingly scarce. In the abridged version of the same text, published in *Conjectures and Refutations*, they do not even appear (Popper 1962, pp. 449–462). The weight of his reasoning falls on one argument: holistic intuition, deemed specific to the social sciences by apologists of Verstehen (or understanding). According to proponents of "anti-naturalism," methods in the physical sciences are not applicable because one cannot proceed atomistically in sociology: "The social group is *more* than the mere sum total of its members" (Popper 1957, pp. 17–19). Groups have a character that goes beyond individuals, based on traditions, institutions, and the like. In physics, structures are mere "constellations": the sum of the parts plus the geometric configuration of those parts. The natural sciences have a method based on causal explanation, induction, and quantification, because they are all combinatorial arrangements (Ibid., Chap. 8, p. 20).

The social sciences, according to the apologists of Verstehen, have a specific method or access to the human world: they work through sympathetic imagination, grasping purposes, meanings, and qualitative aspects. The method of understanding is presented in three variants: first, the rational or teleological model—actions are in accordance with their ends, whether rational or irrational, and aim at achieving advantages, whether real or imagined; second, a situational variant—adding to understanding in accordance with a means-ends scheme the knowledge of the context or situation and confers or modifies the "meanings" of decisions; and third, a variant that goes even further and claims that it is also necessary to know the "mind or spirit of an epoch" and to know historical trends. This last variant is the one that is properly "holistic" (Ibid., p. 24; Gellner 1959,

1963). Two anti-naturalist arguments are independently treated: the limitation of theories to a time or social system and the difficulties of experience.

(a) Periods and generalization: periods do not limit the validity of generalizations, but obviously Popper does not believe in the method of generalization through induction, neither in the social sciences nor in physics. However, at the root of this statement is a certain parochialism which "overrates the significance of the somewhat spectacular differences" and minimizes the possibilities of scientific inventiveness (Popper 1957, Chap. 26, p. 98: "theories are prior to observations as well as to experiments"; what is necessary is only that mistakes can refute theories. It is not in fact even necessary to discuss induction in the natural sciences to criticize the historicist assertion that generalization is circumscribed to concrete periods). So also "there seems no reason why we should be unable to frame sociological theories which are important for all social periods" (Ibid., p. 101; Newton did not need to change the world to formulate the law of inertia). Popper also explains that he does not believe that more general social theories can be obtained inductively by the abstraction of regularities, habits, legal proceedings, and so on, distinguishing between *principia media* and axiomata media. Only the postulate of unlimited validity makes it possible to revise theories. The argument is not as good as it seems, because many generalizations in the social sciences are accompanied by mental restrictions and do not explain or clarify their conditions.

(b) Difficulties of experience: the historicist holds that the experimental method cannot be applied, because "we cannot, in the social field, reproduce at will precisely similar experimental conditions" (Ibid., Chap. 25, p. 93). There are clearly differences between sociology and physics, but this assertion is a crass misunderstanding. A physicist also has great difficulty in seeing what type and degree of isolation are necessary in experience. Artificial isolation is not a problem specific to the human sciences; it exists in physics too: "What kind of artificial isolation, if any, is needed" is decided only on the basis of the results of experience (the Hesiod inference; Ibid., p. 100). People hold the mistaken belief that the generalities they observe around them are the universal laws of social life, but a trip abroad is sufficient to contradict this, whence they extract the inference that everything changes (Cf. Popper 1962, Chap. 23).

The difficulties in repeating experiments under equivalent conditions are surmountable, as in physics; everything "depends on the kind of experiment" (Ibid., p. 94) and is impossible to decide a priori. The problem itself is the information we get from experience: the sociologist is deceived with results different from the expectations "formed on the basis of piece-meal experiments"; the historian has difficulty interpreting certain documents, or discovers incorrect interpretations by his predecessors; the anthropologist feels it difficult to "adjust our habits of thought" in the face of unusual realities (Ibid., p. 96). Even these refutations are nuanced. Although the variability of experimental conditions is pervasive, even in the physical world, it is not insurmountable. As long as we do not want to change everything at the same time—hence an incremental approach—the isolation of variables is always very difficult. Nevertheless, he recognizes that a physicist is in many cases better positioned than a social scientist, for "in practice, it is often very difficult for the social scientist to choose and to vary his experimental conditions at will" (Ibid., p. 97), and he is often restricted to thought experiences and historical analyses.

The criticism confirms in part the force of the arguments. More importantly, the argument which he later judges to be capable of conclusively refuting historicism—the modification of the conditions of modification, including those resulting from the novelty of science—seems to confirm that "laws" may vary with the ages. It seems reasonable to assume that the novelty of the periods, if not complete, can be substantial if the conditions for modification change. (Elsewhere Popper says that the effects of foresight on human behavior would be the distinguishing feature of the social sciences.) This form of presenting the theses that he wishes to refute is characteristic of Popper. In accordance with his own hermeneutical rule, he construes the best possible theory of historicism (Jarvie 2001, p. 149; Popper 1962, Vol. I, pp. 215 and 246, n. 45). This is the best procedure for criticizing "erroneous theories," such as generalization, essentialism, intuition, inexact prediction and complexity, or difficulties in quantifying in social sciences.

Historicism exaggerates the differences between the methods of the natural sciences and the methods of the social sciences. The "burning interest in questions of origin" at the heart of historicism is, however, misplaced (Jarvie 1986, p. 70). But, in this case, the construction of the best possible theory leaves force to some of the arguments *against* naturalism. His answers may therefore seem inconclusive. As Popper himself points out, they are not a refutation of historicism. Yet they keep the door open to the use of scientific method in the social sciences.

Could the Problem of the Method of Social Sciences Have Been Developed Differently?

Jarvie addresses himself to Popper's methodological differentiation between natural and social sciences to tackle a problem: "whether in the years that have elapsed since 1943–1944, when *The Poverty of Historicism* was originally published, there have emerged new arguments to challenge what Popper said about the differences between the natural and the social sciences" (Ibid.). He examines three strands of objections: (1) "the argument from meaningfulness" (there are no data not impregnated with meaning in the social realm), (2) the argument from interests (there is a stronger bias in the social sciences than in the natural sciences), and finally (3) the argument from reflexivity as it "has been admitted in the social description of the pursuit of truth, the social sciences are social too, so the social sciences are in the odd position of having to give an account of social processes and social causation that, if comprehensive, must include the processes and causation of the social sciences themselves" (Ibid., p. 85). Jarvie's short answer to his own question is no, there is nothing new under the sun that Popper had not considered.

However, according to the best Popperian tradition, one learns through mistakes and hence the importance of criticism. So, we shall half-heartedly put forward a criticism ourselves: if the problem was formulated differently, as to whether in the years previous to1943–1944 any valid argument challenged what Popper said about the differences between the natural and the social sciences. Through Hayek's development of subjectivism in the social sciences, we will see how the problem could have been developed differently.

Jarvie dedicates a chapter of his *The Republic of Science* (Chap. 3: "The Methodology of Studying Social Institutions") to the early version of *The Poverty of Historicism* (which is chronologically Popper's first book on the social sciences), focusing on three topics: first, Popper's ideas on the emergence of institutions, including scientific ones; second, his ideas about individuals and their role in science; and finally, Popper's conception of rationality and the scope for objectivity and testability in the human sciences. According to Jarvie, two of the central ideas of the Austrian School that Popper accepted are the theory of the social origin of institutions and methodological individualism (Hacohen (2002) shows how Popper's views on these topics can be traced to Carl Menger and the Austrian School). Popper shares these two ideas in common with Hayek; however, he minimizes a third idea that is central to Hayek: the *subjectivism* of the social sciences.

Scientism, according to Hayek, is the source of a triple error: objectivism, collectivism, and historicism. Like Popper, he sees historicism as the root of totalitarianism, and his hope is that the exposure of these errors will remove collectivism and totalitarianism from its intellectual foundation. Hayek focuses on three errors which seem to be the specific features of the method of the social sciences: subjectivism, methodological individualism, and the possibility of constructing a "nonhistorical" social theory. Despite sharp differences between them, Popper agrees with all but the first one.

(a) *Subjectivism and objectivism*: According to Hayek, the social sciences, in the strict sense, are interested only in conscious or reflective actions, where there is a choice between different courses of action. Although it is also possible to describe situations from the outside, doing so as reactions to stimuli would be to "confine ourselves to less than we know" (Hayek 1964, p. 26). We expect certain patterns of behavior from others and take it for granted that others react in the same way, because they have learned to classify things in a similar way. Most objects of human or social action are not objective facts, nor can they be described in physical terms "irrespective of what people think about them" (Ibid., p. 27). A logical analysis of the concepts of social sciences always shows at least three terms as part of each concept: the person who acts, the desired effect, a "thing" in the vulgar sense. In practice, human behavior is expressed in terms of *purposes*. Furthermore, not only the reference to a goal is necessary, but the concepts used in the social sciences contain *no reference* at all to physical substances, forms, or attributes—no comprehensive definition of a class of human creations is able to include them—such that the definition is made only "in terms of mental attitudes of men towards the things" (Ibid.). What is specific about the terminology of history and the social sciences is not abstraction—all concepts are abstractions—but the abstraction of all aspects or physical properties.

(b) *Methodological individualism and collectivism*: In opposition to Auguste Comte, Hayek also observes that, unlike what often happens in physics, in the investigation of the human world, the most familiar elements are individuals' attitudes. This is why the method of the social sciences is synthetic or "compositive": social scientists construct concepts and models from subjective phenomena. The collectivist or holistic error, by contrast, tries to treat wholes as objects with their own laws. Now these totalities (capitalism, working class, democracy, nation) are not realities like a stone. They are hypothetical constructs, provisional theories which exist only in the popular mind. Confusion often ends in conceptual realism which attributes its own reality to a pseudo-collective spirit. Attempts to give an objective description of these collectives means in practice pretending to know less than is known. Moreover, sociological macroscopic observations do not capture interrelations, for even statistics always begin by eliding the links between phenomena to be able to quantify them (Ibid.).

(c) *Social theory and historicism*: The compositional method is therefore the specific theoretical mode of the human sciences—at the origin of the theories are immediately known subjective phenomena. This is because it is not possible to carry out controlled experiments in the human domain. Theories do not refer to particular constellations of facts, as observed in history, but to abstract phenomena. The historian constantly resorts to theories and concepts from different theoretical spheres, most of which are very simple, embedded in (sociological) concepts such as "battle," "market," or "city" (Ibid., examples in p. 72).

Is the *Methodological* Distinction Between Natural and Social Science More Pronounced?

As we can see, Hayek makes a case for a more pronounced distinction between the methods of natural and social science, and therefore Hayek's argument from subjectivity should be taken more seriously. Despite the qualifications pointed out, Popper does not doubt the fallacy of the arguments that oppose the application of the scientific method to the human world. If we look at the present-day panorama of social sciences, it is clear that some branches do correspond to Popper's description of a non-historicist approach to human phenomena, say, microeconomics, formal linguistics, and public choice theory; there are general sociological laws or hypotheses. Sociological laws or hypotheses are similar to physical theories, and Popper enumerates a few examples: full employment without inflation is not possible, the concentrated interests of producers prevail over those of consumers, all power raises occasions of abuse and Acton's law (Popper 1957, pp. 62–3). We will find they are innumerable, if only we "we look for them among the more platitudinous rather than among the more abstract laws" (Ibid.). Many of them can be tested and refuted.

Could it be that Popper and Jarvie underestimate the differences in method between the natural and social sciences? Might Hayek not be right in stressing that the effort to give up *subjective opinions* and arrive at *objective facts* does not apply in the human sciences, because the purpose of these sciences is precisely "opinions" that guide human activity? At the same time, the "opinions" that are the subject of human study are not those of the researcher, but those of the agent, so that it can also be said that they are not subjective because "they are independent of the particular observer." Here we find again Max Weber's proposal about retrieving the subjective meaning of a social agent (cf. Weber 1965; see Colen 2017).

In spite of this, "beliefs or opinions held by particular people, beliefs which as such are our data, irrespective of whether they are true or false," are data of a different nature compared to natural facts, because not only do they not feature physical attributes, but also, to top it all, seeing that we cannot observe others' minds directly, we can only perceive what they are doing because we have "a mind similar to theirs" (Hayek 1964, p. 28). This similarity of "the structure of the mind" does not for this reason stop being an empirical fact, which is confirmed both by the possibility of communication and by the results of research: if sensations do not correspond to physical properties—qualities disappear from the scientific world—they continue to be part of the human mind. How does Hayek justify using an ambiguous expression? The term "subjective" also stresses that, in spite of the essential structure of the mind being common, what makes communication, categories, and opinions possible "will yet be different and often conflicting in many respects" (Ibid., p. 29); otherwise, the distinction between objective and subjective could in practice be disregarded, because the knowledge, concrete and circumstantial, that guides human action rarely seems organized as if it were a consistent and coherent corpus. "It only exists in the dispersed, incomplete, and inconsistent form in which it appears in many individual minds" (Ibid., p. 30). This dispersion and imperfection is a basic fact for the social sciences.

To sum up, tools, sentences, drugs, cosmetics, food, and money are not objective facts but must be defined on the basis of the way in which one thinks. There is no physical distinction between an exchange in the market and a magical ritual (gift and counter-gift). Only when the meaning of actions can be seen isit possible to classify and explain them: "to subsume them under rules which connect similar situations with similar actions, are bound to fail" (Ibid., p. 31). The substantiation of this methodological detail in social studies has been especially useful in economics and perhaps in linguistics. Ricardo's law of rent serves Hayek as an example of a general statement which is joined to an empirical proposition that varies with circumstances. It has a genuine explanatory value provided that the circumstances that are its conditions of validity are verified.

Popper's approach might in fact open a path to some of these insights. His distinction between the theoretical and historical sciences (within the ambit of the social sciences) is predicated on the belief that the former aim at establishing general trends, while the latter are concerned with specific instances (Popper 1957, Chap. 30). Beyond simple antiquarian interest, the utility of specific instances lies in the opportunities they offer us to challenge and falsify our general hypotheses. Indeed, oftentimes they attract our interest precisely because they deviate from our understanding of general behavior. It is in these deviations that we must have recourse either to other general forms of behavior that can account for the deviation or to the beliefs and intentions of the actors in that specific instance. If the good scientists constantly launch challenges against their own hypotheses, then they are compelled to seek out the abnormal cases which are comprehensible only through Verstehen. In this respect, Hayek's subjectivism does indeed distinguish the method of the social sciences from the method of natural sciences. A concern for Verstehen and the specificity of each moment in time is what we encounter in both history and literature.

The Republic of Science and the Social World at Large

In *The Republic of Science* (2001) Jarvie argues that, far from ignoring the social dimension of science, Popper, almost from the very beginning, presents a strong sociological program and even suggests, implicitly, a kind of "proto-constitution of science." This may surprise many since the conventional interpretation of Popper's thought avers that he was, above all, a philosopher concerned with the methodology of science, with interests extending from physics to biology, and merely took a detour, so to say, in *The Poverty of Historicism* and *The Open Society and Its Enemies* (2009), the latter described as his contribution to the war effort (Popper 2006). Contrary to this received wisdom, Jarvie convincingly argues that an (decisive) element of Popper's legacy is his showcasing, from 1934 onward of the role that methodological conventions or social "rules of the game" play in science. In this way, Jarvie shows that Popper anticipates the social shift in the philosophy of science by decades: "thinking socially (rather than logically or psychologically) is central to Popper's philosophical enterprise beginning with *The Logic of Scientific Discovery*, continuing for his ten most creative years, and emerging sporadically after that. Popper's consistent ability to think socially also does much to account for his originality, since it is hard to do and its difficulty is attested by how often readers and critics of Popper do not grasp that this is what he is doing" (Jarvie 2001, p. 21). For Jarvie, Popper had much to say about a scientist's (social) conduct. Although apparently only "logical" or epistemological, the core of *Logik der Forschung* implies an institutional setting. Moreover, Popper's concern with methods in the human sciences (explicitly excluding pseudosciences, such as Marxist sociology and psychoanalysis) precedes his turn to the natural sciences. Popper's approach to science is thus a mirror of his political philosophy: just as the function of the philosophy of science is to criticize the "rules of the game" of the natural sciences, so is the function of moral and political philosophy is to question conventions, customs, traditions, and institutions.

Jarvie expands Popper's proposals and makes it explicit that the rules of scientific procedure must be *socially* designed "in such a way that they do not protect any statement in science against falsification" (Ibid.). He lists a set of articles of the proto-constitution of science which, he acknowledges, were very abstract in his first book: Popper later corrected and supplemented them with a theory of institutions. What Jarvie calls a "Republic of Science," the regime which is established with a "proto-constitution" is later promulgated with an enriched theory of institutions. It is impossible to do justice here to the originality of this reading, and it may be preferable to point the reader to Jarvie's work itself. Jarvie, however, doubts that the workings of the "community of science" can be extended to society as a whole. His doubts are well founded. The world of science is not that of common sense; but common sense is an experimental datum in the human realm. It provides enough of an image for humans to orient themselves in daily life but is an imperfect tool that needs improving. Science is engaged in revising the image of the world, and as a real fact it implies that all thinking beings possess a similar image of this world. Humans see the world by way of sensations and organized concepts in a "mental structure common to all" (Hayek 1964, p. 23). Until science completes its work and explains intellectual processes, it is necessary to use the facts of the mind to explain human actions, because these are determined by the views and concepts he possesses, and through these mental facts, man builds up another world of which the individual becomes a part (Ibid., p. 24).

Formal science is not the duplicate of events experienced, whether in the social sciences or the natural ones. It is sufficient to consider the difficulties we would face and the quantity of physical and astronomical laws or principles of thermodynamics we would need to scientifically describe an event as simple as "It was nightfall and I turned on the heating and the light on my desk." Science, physical or human, explains some aspects, cut to fit its laws, which it takes from events. There is certainly space for introducing progressively more complex models, but the joining of the experienced and the formal will only take place when chemistry substitutes cooks in foreseeing the flavor of a dish. This does not constitute a problem for the natural sciences which are always unfinished and are concerned with adapting the image of the world to the facts. But the same cannot be said about the sciences that try to explain *and guide* human actions, *here and now*. A different task, for which the same methods may not be suitable, is at stake (example taken form Veyne 1996).

Conclusion: Can a "Republic of Letters" Play a Role?

To put it somewhat differently: why is there no room for building a "Republic of Letters" along with the Republic of Science? As Jarvie vividly explains, "Popper presents society as a learning machine, and the learning machine as a Socratic seminar writ large" (Jarvie 2001, p. 6). He avers that "there are serious deficiencies in holding up the Socratic seminar as a model for life in society in general, and for scientific work in particular" (Ibid.). This last problem, which goes beyond Popper's legacy, in letter if not in spirit, is probably *the* single most important question faced by scientists and philosophers of science. "The republic of science Popper envisions is susceptible to corruptions of power just as is the broader body politic, and its citizens need the link of checks on power that Popper demanded on governments. More troubling: the central Socratic and scientific value of truth conflicts at times with other values important for social life" (Ibid.).

This may also surprise many, since the successful part of the modern project, since Descartes, Bacon, and Hobbes, has obviously been science, even if the benefits of the effects of its application in areas like genetic improvement, nuclear weapons, climate change, pollution, or even the financial system seem questionable. But science's prestige remains untouched, and it is to science, not philosophy, that we look for all the answers to current problems. Philosophy does not seem to produce cumulative knowledge but is instead always entangled in perennial issues. If there is a "truth" in matters of philosophy or metaphysics, it seems inevitable to conclude that the history of philosophy is a history of mistakes and failed attempts.

Bearing in mind Jarvie's contentions with the extension of the republic of science to society, we might conclude by adding another contention that would recommend a new republic of social sciences or a republic of letters as a true Socratic seminar. It is not just that a Socratic seminar writ large, characterized by an ethos of critical rationality, would run up against problems of a lack of checks on power or the conflict of values, such as that between the Socratic search for truth and those values essential to maintaining a political community. It is also that the Socratic search for truth is characterized by a concern for the ends that motivate human behavior. As such, a Socratic seminar would include rigorous efforts to falsify standing hypotheses, but it would also not shy away from Socrates' practice of asking questions of "essence": "what is F?", F being piety, courage, wisdom, justice, and so on. Perhaps we cannot and should not ignore these questions that all societies have put to themselves through the ages, questions that may produce conclusions that are consequential and falsifiable, even if they may not belong within the Republic of Science.

References

Brudny, Michelle-Irère. 2002. *Karl Popper, un philosophe heureux. Essai de Biographie Intellectuelle*. Paris: Grasset.

Colen, Jose A. 2010. *Futuro do Político, Passado do Historiador*. Lisboa: Moinho Velho.

———. 2017. *Facts and Values: A Conversation*. London: Plusprint.

Gellner, Ernest. 1959. Holism versus Individualism in History and Sociology. In *Theories of History*, ed. Patrick Gardiner, 604–622.

———. 1963. Nature and Society in Social Anthropology. *Philosophy of Science* 30: 236–251.

Hacohen, Malachi H. 2002. *Karl Popper – The Formative Years, 1902–1945*. Cambridge: Cambridge University Press.

Hayek, Friedrich A. 1964. *The Counter-Revolution of Science: Studies on the Abuse of Reason*. London: The Free Press.

Jarvie, Ian C. 1986. *Thinking about Society: Theory and Practice*. Dordrecht, Holland: D. Reidel Publishing Company.

———. 2001. *The Republic of Science. The Emergence of Popper's Social Science. 1934–1945*. Amsterdam: Rodopi.
Popper, Karl R. 1962. *Conjectures and Refutations: The Growth of Scientific Knowledge*. New York and London: Basic Books.
———. 1964 [1957]. *The Poverty of Historicism*. New York and Evanston: Harper.
———. 2006 [1974]. *Unended Quest: An Intellectual Autobiography*. London and New York: Routledge.
———. 2009 [1945]. *The Open Society and Its Enemies*. London: Routledge.
Shearmur, Jeremy. 1996. *The Political Thought of Karl Popper*. London and New York: Routledge.
Veyne, Paul. 1996 [1971]. *Comment on écrit l'histoire*. Paris: Seuil.
Weber, Max. 1965 [1922]. *Essais sur la théorie de la science*. Trans. Julien Freund. Paris: Plon.

CHAPTER 16

Karl Popper, the Open Society, and the Cosmopolitan Democratic Empire

Malachi Haim Hacohen

Introduction

"I am reading with great interest your dissertation and hope it will become a book," wrote me Ian in an October 1994 email, our first communication. "Not that I agree with everything in it, far from it," he added. Then, he proceeded to invite me to contribute to a special issue of *Philosophy of the Social Sciences*, "The Open Society After 50 Years." This was the beginning of a decade long correspondence on Popper and his philosophy. We met in person on several occasions: I invited Ian, in 2000, for a Vienna "rehearsal" of "Scientific World Conceptions," a summer course that has since become the flagship program of the University of Vienna Summer School. We then attended together the HOPOS conference. We met again at the Karl Popper Centenary Congress in Vienna two years later. My essays on Karl Popper (Hacohen 1996, 2006) outline the Popper phase in my life. Ian was an integral part of it, someone with whom I could discuss Popper's biography, politics, and sociology of science. Ian knew Popper well, both as his student in the 1960s and as a scholar, but he charted his

M. H. Hacohen (✉)
Department of History, Political Science, and Religion,
The Kenan Institute, Duke University, Durham, NC, USA
e-mail: mhacohen@duke.edu

R. Sassower, N. Laor (eds.), *The Impact of Critical Rationalism*,
https://doi.org/10.1007/978-3-319-90826-7_16

own course in the social sciences. He was patient with our differences, and unfailingly supportive with my book on Karl Popper, an ideal interlocutor and mentor.

Ian and I never reached an agreement on Popper's Jewish politics. I now understand that this was less because he had a different position from mine on the Jewish Question and more because, like Popper, Ian is a secular philosopher. He is tolerant of religion, but religious tradition does not speak to him the way it does to me. As I turned to *Jacob & Esau*, a monograph on two millennia of Jewish-Christian relations, our friendship remained but our correspondence dwindled (2018). This chapter represents the bond that still ties us. In *Jacob & Esau*, Popper's *Open Society* represents the culmination of an Austrian imperial tradition that enjoyed both Jewish and socialist support. The Jews, who regarded the Austro-Hungarian Monarchy as a bulwark against antisemitism, were famous as imperial patriots. Until the end of WWI, the socialists, too, supported the Empire and proposed federalist democratic reforms to modernize it. In his New Zealand exile, Popper dreamed of a cosmopolitan democratic empire that would contain ethnonationalism and accommodate heterodox Jews and socialists. In my earlier work (2000), I thought Popper's views on nationalism were out of sync with reality; now I see them as prescient. I hope Ian will be receptive to my *mea culpa*. This chapter may represent a meeting of our minds and a reaffirmation of our bond of friendship.

I shall begin by evoking a prominent anthropologist, known to Ian from his time at the London School of Economics, Ernest Gellner. Gellner spoke of nostalgia for the Austrian Empire as the product of pariah liberalism (1998, pp. 30–36). Popper, Gellner's mentor and lifelong nemesis, provided his model for pariah liberalism. Gellner himself was a pariah liberal, but, unlike Popper, he also had warm feelings for Czech nationalism and never felt quite as much a homeless cosmopolitan (Hall 2010). Yet, in its time, the Empire was least beloved by liberals, even if Jewish. As a pro-imperial Austrian progressive, Popper was more of an exception than the rule. He represented an *acte final* in the curious coalescence of empire, socialism, and Jews (Hacohen 2018, Chap. 7). His rediscovery of empire during WWII was momentous for progressive and socialist discourse. Holding Austria-Hungary's dissolution to be a major tragedy, and nationalism and the nation-state to be the scourge of modernity, and thus responsible for the catastrophe that had sent him fleeing from Europe, Popper envisioned a cosmopolitan democratic empire. No Austrian Socialist ever dared dream of such an empire, because few had made as clean a break with German nationalism or were as critical of Marxism as he was. Popper called his empire the Open Society, and it shaped the post-WWII Cold War liberal imagination. The coalescence of empire, socialism, and Jews helped form the postwar world even after all three had vanished from Europe.

A stranger in his homeland even before he emigrated in 1937, Popper has become, in recent decades, something of an Austrian national philosopher. Born, raised, and buried in Vienna, he is still known primarily as a Western intellectual, an anti-communist prophet of Cold War liberalism. Yet he was the foremost philosopher to refashion the progressive Viennese legacy into postwar transatlantic liberalism. His cosmopolitan empire emerged from the Austrian Jewish experience of national exclusion, yet provided the opposite response to that of the Jewish nationalists. Whereas they were looking for a pluralist empire to create national autonomy, Popper delighted in empire's ability to break down national boundaries, initiate a cross-cultural dialogue, and give rise to cosmopolitanism. His cosmopolitan democratic empire offered opportunities and represented problems not encountered in earlier discourses. Having emerged during WWII, when the path of Austrian socialism seemed blocked, it disappeared in postwar culture. It is time to recover it.

Viennese Progressivism and Cosmopolitanism

The fortunes of the Popper family tell the story of the mercurial rise and decline of late imperial Habsburg Jewry. Karl's father came from Bohemia and his maternal grandparents from Silesia and Hungary. In Vienna, they adopted German culture, sent their children to German schools, and became part of Vienna's professional and commercial elite (Rozenblit 1983; Beller 1989). Simon Popper became the legal partner of Vienna's last liberal mayor, Raimund Grübl, took over the firm in 1896, and the family moved into a huge apartment with adjoining offices, across from the St. Stephen's Cathedral (Popper 1976, pp. 8–10). He married "up": Popper's mother, Jenny Schiff, was daughter to a family of the Viennese high bourgeoisie. Entrepreneur Max Schiff had come from Breslau, made a small fortune, and had an apartment in the ninth district and a villa on the outskirts of Vienna. He also became a benefactor of the arts. The Popper household embodied the ideals of *Besitz* (property), *Recht* (law), and *Kultur* (culture) (Ibid., pp. 53, 82; see also Schorske 1980).

In 1900 Simon and Jenny Popper renounced their membership in the Jewish community, converted to Lutheranism, espoused liberal anti-clericalism, preferring the *Aufklärung*'s religion. Vienna had the highest conversion rate of any European urban center, and Lutheranism was the religion of choice for upper-class Jewish converts. Assimilated Jews remained a small minority, however, and neither acculturation nor religious conversion broke the barriers of ethnicity. Upper middle-class assimilated and non-assimilated Jews belonged to the same social networks. Progressive Jewish intellectuals built bridges to non-antisemitic secular Austrians, but both groups remained marginal within their own cultures. The Poppers spent much of their life in the company of other Jews.

The progressive intelligentsia rebelled against the social conservatism of mainstream liberalism, and sought a dialogue with socialists. They organized a political party (*Sozialpolitische Partei*), but it remained small, running against the twin obstacles of Catholicism and antisemitism. The progressives increasingly channeled their efforts into a large network of associations for educational reform, social welfare, and economic planning (Holleis 1978; Belke 1978, pp. 5–56; Fuchs 1949, pp. 133–162; Boyer 1978). The young Popper's social and intellectual milieu was secular and politically radical, trusting in social reform, popular education, and technological progress. It reflected a late-enlightenment spirit (*Spätaufklärung*) (Stadler 1981).

Ethnonationalism was *Spätaufklärung*'s greatest enemy, but the progressives failed to understand it, underestimated its danger, and responded ambivalently to it. Their ranks included pacifists and federalists, but also German nationalists. They fought antisemitism that offended their humanity and excluded their Jewish members from the German nation, but they could see no harm in expanding the German cultural sphere in Central Europe and regarded Slavic nationalism as reactionary. The *Sozialpolitiker* declined to endorse federalism, or Premier Casimir Badeni's language ordinances. Like the Socialists, they piously suggested that a democratic electoral reform would facilitate agreement among the nationalities. An enlightened administration, encouraging economic development in German areas, could resolve the Czech–German conflict (Hainisch 1892; Wittelshöfer 1894). Most of them refused to implicate German nationalism with antisemitism. They believed that antisemitism was rooted in religious prejudice—amply demonstrated by the Christian Social Party—and secular education was its proper antidote. Clericalism, not nationalism, was their major enemy (Boyer 1995, Chap. 4). They contested antisemitic rhetoric with German *Aufklärung* rather than with imperial pluralism.

The Popper family represented the cosmopolitan pacifist pole on the progressive spectrum. Simon Popper was master of the leading freemason's lodge *Humanitas*, and relatives and friends were identified with the Austrian Peace Movement (Popper 1976, pp. 11, 13–14). Unlike the Socialists, neither the Freemasons nor the pacifists drew plans for imperial reform, but the Peace Movement gave rise to the foremost *fin-de-siècle* imperial cosmopolitan project, Alfred Fried's (1864–1921) proposal for a federalist Europe. Entertaining a materialist evolutionist view of human development, Fried regarded the nation-state as a mere stage in the growth of international government, and he expected it to encompass eventually the entire globe. Global government's growth would be gradual, a matter of scientific necessity. To advance it, he proposed a federalist European union, extending to imperial collaboration overseas. As Katherine Sorrels aptly observes (2016, pp. 65–100), all pre-WWII Pan-European projects were vested in continued European global domination, but, in contrast with racial supremacist schemes, Fried envisioned progress toward universal egalitarian humanity. He was the prime example of progressive Viennese cosmopolitanism prior to Popper.

For the rest, the progressives' grand designs for a future society had little use for Austria-Hungary and, instead, tended toward utopia. Imagined progressive communities were modern, technologically advanced, and socially engineered, but they seemed context-less. Neither Theodor Hertzka's *Freiland* (1890, free land) nor Anton Menger's *Arbeitsstaat* (1904, labor-state) nor Josef Popper-Lynkeus' *Nährarmee* (1912, nutrition army) was tailored to the Monarchy. They may have surreptitiously expressed supra-nationalism, but they also reflected the progressives' inability to negotiate imperial problems: They divested their utopias of any national attribute, and understood the requirements of a good social order to be a matter of scientific management, not cultural difference. They confounded *Aufklärung* and *Deutschtum*, but, for most, it seems, *Deutschtum* was an instrument for realizing *Aufklärung*, rather than the opposite. They were less nationalists in a cosmopolitan guise than cosmopolitans in a German guise, and in despair (Beller 1996).

Like the Socialists, the progressives denied that Jews were a nationality. Striving for recognition as German Austrians, they sought to strip religion and ethnicity of significance—their own first and foremost. Their utopias were commonwealths free of religious superstition and ethnic prejudice, where they would finally find a home: No one would probe their ethnic origins or challenge their claims to be German. The progressive Jewish intelligentsia could gain nothing from ethnonationalism: They were "a class … which claims no particular [national] right … but can evoke only a human title [and] cannot emancipate itself without … the complete redemption of humanity" (Marx 1961, Vol. 1, p. 390). To overcome the burden of their own ethnicity, they needed to dissolve all ethnicity and recover universal humanity. Cosmopolitanism represented the response of ethnopolitics' losers.

Popper would spend much of his life refashioning progressive philosophy and politics. He purged Viennese progressivism of its ambiguity about German nationalism, relieved Austrian socialism of its German commitments, and highlighted the cosmopolitanism of both. Nowhere was this more evident than in his reintroduction of empire into progressive discourse. Responding to the predicament of German-Jewish identity, Popper rejected both German and Jewish identities in favor of cosmopolitanism and made the democratic empire cosmopolitanism's agent. No one else, progressive or socialist, went this far. Popper remained a permanent exile, a citizen only in an imaginary Open Society. Rejecting *Deutschtum* and dissociating the *Aufklärung* from Germany, Popper vindicated the Enlightenment, but also inherited its dilemmas. Discounting all national, ethnic, and religious identity, he posited universal visions of the Open Society and the democratic empire where none of them counted (1945).

Socialism and the Democratic Empire

"The breakdown of the Austrian Empire and the aftermath of the First World War … destroyed the world in which I had grown up," wrote Popper (1976, p. 32). Following the dissolution of the Empire in the fall of 1918, he left the *Realgymnasium* to participate in the Austrian Revolution. He first joined the Socialist students, then the Communists, but he was not one to accept party discipline and dogma, and, after the failed Communist *Putsch* of June 1919, he dissociated himself from the party. His rejection of Communism was, however, a prolonged process, stretching over years (not months, as he claimed), and leading him eventually to withdraw from social engagements altogether and focus on his academic career (Ibid., pp. 7–8, 10, 31–41, 53–5, 71, and 197 n. 2). Still, he remained a Socialist and left Vienna in January 1937 considering himself a member of the now banned party.

Popper was, however, a socialist dissenter, and he was developing a critique of Marxism and the Austrian Socialists. Already in 1924, he was suggesting that Eduard Bernstein's revisionism and Carl Menger's marginalism provided an alternative to Austro-Marxism. In "Toward a Philosophy of the Homeland (*Heimat*) Idea" (1927), he viewed Immanuel Kant, and not Marx, as showing the way to socialism and internationalism. Noting the amorphous character of homeland (*Heimat*), he expressed anxiety lest it foreclose new experiences as "foreign" and purged the concept of romantic nationalist meanings. *Heimat* established a "naturally given primitive cultural community" but the nation was a legal association, and law, and not cultural heritage, set national boundaries. A universal ethic must shape national legal codes, making good citizenship compatible with internationalism. Education should cultivate respect for law, a sense of justice, and a critical awareness of social inequities. It should not foster patriotism, or love of *Heimat*. "From good Germans to good cosmopolitans" (Ibid.), he quoted Eduard Burger, a noted socialist school reformer. He offered the Socialists no help in appropriating *Heimat* from the Right; instead, he sought to free Socialism from the German nationalist *Heimat* and return them to internationalism.

Popper and his future wife were among the crowd in front of the *Justizpalast* on Bloody Friday, July 15, 1927. They watched incredulously as the police opened fire on "peaceful and unarmed social democratic workers and bystanders. We were lucky to escape" (draft of 1976, p. 135). He thought that the police's attack was unprovoked, but, all the same, blamed the socialist leaders for the massacre. Their "suicidal" policies gave the government an opportunity to use violence. He understood well the event's historical significance. His call in *The Open Society* for a steadfast defense of democracy encapsulated his policy proposals to the Austrian Socialists. He thought that the government's commitment to democracy was shaky and the fascist threat real. The Socialists needed to contain their rhetoric so as not to provoke a *coup*, but failing that, they had to defend democracy by force. The two goals may have been in conflict, but, already as a young man, Popper saw the situation clearly, and his socialism was *sui generis*.

In the scientific philosophy of the Vienna Circle, Popper found—for all his lifelong confrontation with it—the legacy of the Enlightenment and of *fin-de-siècle* progressivism. The Circle was the fountainhead of the logical positivist movement that, during the interwar period, developed an organizational network in Central Europe's urban centers: Vienna, Berlin, Prague, Warsaw, Budapest, Lwów, and Bratislava. It had disciples throughout Europe and North America. The Circle sought to apply recent advances in logic, mathematics, and scientific theory to philosophy. Many members were deeply influenced by Ludwig Wittgenstein's *Tractatus Logico-Philosophicus* (1922) and declared war on traditional philosophy, especially metaphysics. Among the more famous were Moritz Schlick, Otto Neurath, and Rudolf Carnap. Popper's relationship to the Circle was problematic. He developed his philosophy of science in critical dialogue with theirs, but within a framework foreign to positivism, that of the marginal Kantian tradition leading from Jakob Fries to Leonard Nelson. Recognizing his originality, the Circle provided him with opportunities that eventually made him famous, but the disjunctions between their positivism and his Kantianism remained a source of constant tension (Popper 1976, pp. 72–90; see also Feigl 1969; Hacohen 2000, Chaps. 5 and 6; Stadler 1997). All the same, having completed *Logik der Forschung* in 1934 (1959), Popper became an active member of the Circle's network, a "Central European intellectual."

By 1934, however, the Circle's network was living on borrowed time. In Germany it could no longer operate openly. Most members lost their academic positions and emigrated when they could. In Austria they were under attack in the academy and considered suspect by the *Ständestaat*. Proto-fascist regimes were harassing them in other East-Central European countries, too. None of the Circle's annual congresses for scientific philosophy, beginning in 1935, could take place in Central Europe. Instead, they were held in Paris, Copenhagen, Cambridge (UK), and Cambridge (Massachusetts). They solidified British and American interest in scientific philosophy and facilitated the circle members' migration. Within a few years, most of them left Central Europe for England and the US.

After the failed Nazi *coup* of July 25, 1934, Popper thought that a Nazi take-over and a German invasion of Austria were merely a matter of time. He had no chance of an academic appointment and was searching desperately for a way out of Austria. After much travail, he accepted an offer from Canterbury College in Christchurch, New Zealand. In January 1937, he left Austria forever, leaving behind friends and relatives, including his sick mother and a sister (Hacohen 1996, pp. 455–57). As German troops overran Europe, and the Japanese advanced down the Pacific Ocean, he worked feverishly to complete *The Open Society*, thinking at times that he might be writing civilization's testament (Simkin 1993). The book was his contribution to the war effort, and he invested it with a fighting soldier's zeal. *The Open Society* explained the triumph of fascism, the surrender of Central European democracies, the issues at stake in the war, and the principles for social reconstruction afterwards.

The Open Society consisted of three controversial critiques of Plato, Hegel, and Marx that traced modern totalitarianism to the "intellectual leaders of humankind" (1945, Vol. I, p. 1). Popper opined that the fascist drive to return to "tribal" society fed on anxieties arising from rapid changes in open societies. Plato conspired to halt Athens' economic development and substitute an authoritarian regime for democracy. Hegel twisted progressive philosophy and politics, so that they served Prussian "tribal nationalism." Marx was essentially a progressive democrat, but his scientific pretensions concerning historical inevitability weakened socialist resolve to confront fascism. Fascism's rise and socialism's eventual triumph were, thought socialists, historically predetermined. They fought fascism half-heartedly, leaving liberal democracy defenseless. The fascists were quick to destroy democracy.

Postwar reconstruction of an open Europe was Popper's project. His *Open Society* was the assimilated Jewish philosopher's cosmopolitan homeland, an imagined community that would end his national exclusion and cultural marginalization. It was a society that paid no attention to one's ethnic origins or religious and national affiliation. It was democratic, but it resembled an empire and not a nation-state. Empires provided a necessary transition between tribal nationalism and cosmopolitanism, and were the only effective remedy to nationalism. Imperialism lowered barriers among ethnic groups, permitted enlightenment to penetrate into ghettoes, and forced closed communities open. Popper was willing to see an imperial metropolis temporarily gain at the expense of the peripheries for the sake of an eventual dissolution of the boundaries between center and margins. He willingly took the risk of imperial hegemony of a (German or Hungarian) *Staatsnation* in the hope that it would promote *Aufklärung* rather than racial *Deutschtum*.

The Austrian legacy explains Popper's imperial idealism. He thought that German nationalism and Austrian imperialism were at odds, and that *Aufklärung* and empire were concordant. His charitable view of the British Empire reinforced his imperialist bias. The British Commonwealth, he thought, was the largest free community on earth. He was little aware, if at all, of the British vogue of progressive imperialism, but the Commonwealth first extended hospitality to him in its metropolis, England, then provided him with a refuge in New Zealand during the war (Zimmern 1934). German totalitarianism and British democracy were at war. If the democratic cosmopolitan empire could be saved, nationalism would subside and enlightenment would spread. *The Open Society* represented the last metamorphosis of late imperial Austria, projected onto the British Empire.

The democratic cosmopolitan empire dominated *The Open Society*. The history of humanity was a struggle between progressive imperialism and reactionary nationalism. Classical Athens was the first open society, the first to break tribalism's chains, overcome myth, magic, and custom, and found politics on *logos* and law. Secular, commercial, democratic, and cosmopolitan, it embodied the progressive imperial ideal, but, eventually, succumbed to totalitarian Sparta (Hacohen 2000, pp. 410–6; 2002; compare: Zimmern 1911). Later, Alexander the Great and the Roman Empire advanced universalism: "From Alexander onward, all the civilized states of Europe and Asia were empires, embracing populations of infinitely mixed origin. European civilization [has] remained international or, more precisely, intertribal ever since" (1945, Vol. 2, p. 48). In the nineteenth century, tribal nationalism reared its ugly head once again. Hegel and German intellectuals promoted Prussian nationalism and, within a century, Central Europe succumbed to German barbarism. Nationalism must be reined in, if postwar Europe is to be an Open Society. An international legal order would best accomplish such a task (Ibid., Vol. 1, Chap. 10; Vol. 2, Chap. 12).

Central European conceptions of nationality and imperialism shaped Popper's account. Imperial Austria made *Volksstamm* (ethnicity) the basis of claims for *Nationalität* and cultural autonomy. Popper spoke therefore of "tribal nationalism" (ethnonationalism). There were no good and bad nationalisms: he would not second a cultural *Deutschtum* any more than an ethnic one. The only nation he recognized was a political one, founded on the French Revolution's concept of citizenship, and he subjected it, too, to international law. Almost alone among progressives and socialists of his generation, Popper challenged national self-determination and the nation-state. Nationalities did not really exist. German thinkers from Herder to Fichte to Hegel invented them to serve the interests of reactionary states: "The idea that there exist natural units like nations or linguistic or racial groups is entirely fictitious. … The principle of the national state … owes its popularity solely to the fact that it appeals to tribal instincts" (1945, Vol. 1, Chap. 9, n. 7). Moreover, "none of the theories, which maintain that a nation is united by common origin, or a common language, or a common history, is acceptable, or applicable in practice. The principle of the national state … is a myth. It is an irrational, a romantic and Utopian dream" (Ibid., Vol. 2, p. 49).

Wilson's and Masaryk's well-meant effort to apply national self-determination consistently throughout Central Europe—"one of the most mixed of all the thoroughly mixed regions of Europe"—was an incredible folly that brought about the failure of the Versailles Treaty, said Popper: "An international federation in the Danube basin might have prevented much" (Ibid., Vol. 2, Chap. 12, n. 53). National Socialism represented the nation-state's culmination. So did the tragedy of Bosnia, avowed Popper in his 1994 Prague speech. If national self-determination did not lose its authority, post-communist Central Europe could also fall prey to ethnic terrorism. The only remedy was abandoning self-determination, recognizing state boundaries as conventional, sanctioning the status quo, and establishing an armed international organization to guarantee peace. National identities were false, reactionary, and utopian. Individual, imperial, and cosmopolitan identities were true, progressive, and possible (Ibid., Vol. 1, Chap. 5, n. 13, and Chap. 6, n. 44; also Popper 1963, p. 182; 1992, pp. 118–21; 1994a, pp. 185–7; 1994b).

To his contemporaries, Popper's view of nationalism as a curiously successful intelligentsia's fraud, a catastrophe brought upon Central Europe by treacherous intellectuals—*la trahison de clercs par excellence*—seemed incredible. (I, too, was incredulous about Popper's theory of nationalism, but I now concede that he had seen deeper than I did.) Recent decades have seen post-national US historians of Central Europe developing a historiography that sustains precisely such a view (Brubaker 1996; King 2002; Judson 2007). Popper ingeniously deconstructed nationality, pointing out its complex historical formation and diffused character. He never extended, however, the same mode of inquiry to imperialism. The nation-state was historicized and delegitimized; empires went unexamined and were vindicated. His recollection of historical episodes of imperialism was selective. He recalled Alexander's cosmopolitanism, the Napoleonic Code, and Austrian pluralism, not Spanish colonialism, the Middle Passage, and Nazi *Lebensraum*. Imperialism represented cosmopolitanism's possibility: this was enough. The historicity of imperial identities never became an issue. They were divested of historical specificity. They did not emerge from historical identities, but overcame them.

Popper emphasized that if we deconstruct false collectives and get to the individual, we will reach the truly universal. Yet, unlike Austrian imperial identities, his cosmopolitan identity seemed abstract and unreal. Popper himself conceded that "concrete groups"—families, churches, voluntary associations, possibly even ethnic communities—would remain even in an open society. They would continue to fulfill some of the functions that kinship groups had in the "closed society," and so, an absolutely open society was sheer utopia. Indeed, closed groups were essential:

People, Popper said unsympathetically, will "try to satisfy their emotional social needs as well as they can" (1945, Vol. 1, pp. 171, 175). But "emotional social needs" remained foreign to him, implicated with fascism. He never negotiated between the closed and open society, ethnicity and cosmopolitanism, nation and empire or showed their possible convergence in his future cosmopolitan federation.

There was no room for negotiation with fascism. Popper's categorical rejection of the claims of closed communities against the cosmopolitan empire ought to be understood in the context of Central European fascism. He acknowledged "diversity" and assailed efforts to suppress difference. In one of his last public appearances, he acerbically suggested that Germany and France's homogenous populations were due to nationalizing states using "political and educational means [to] suppress minorities or dialects" (1994b). Still, he thought that ethnic and religious differences were insignificant. Imperial pluralism existed, but, unlike universal humanity, it was no cause for celebration, and must not infringe on cosmopolitanism. The first response to the racist argument that those who are different are inferior and cannot be members of the nation is *not* that we ought to respect difference; it is that we are all equally human and entitled to equal rights as citizens.

The Open Society defended this universal vision eloquently. Central Europe ensured that it remained utopian. Ethnonationalism first gave rise to cosmopolitan dreams and then made them impossible. But Popper would not give up. He searched throughout Western history for cosmopolitan moments to provide instruction and encouragement for those fighting fascism. He found Socrates, Pericles, and classical Athens. They became the origin of *The Open Society*. It mattered little that it happened two millennia ago, far from Central Europe. Where the Austrian Empire failed, Athens succeeded. Progressive imperialism triumphed over nationalism and ethnopolitics. Where he found success once, the exile could hope, amidst global ruins, for another.

Cosmopolitanism Without Jews?

Popper presented his views on the Jewish Question as flowing from his cosmopolitanism. Jewish religion and nationality were impediments to cosmopolitanism; hence, assimilation was a moral imperative. In fact, he got the relationship between cosmopolitanism and assimilation wrong. It was precisely the difficulties of assimilation that gave rise to cosmopolitanism. Finding themselves excluded from the nation, Jewish intellectuals imagined cosmopolitan communities that would accept them. Cosmopolitanism became a precondition to assimilation, not as Popper surmised.

An unbridgeable gap separated cosmopolitan dreams from reality, the open society from nationalizing Central Europe. The gap haunted Popper with a vengeance as he tackled the Jewish Question. In stark contrast with his constructionist view of the nation, Popper held an essentialist racial view of the Jews and Jewish history. The Hebrew Bible was, he said, the fountainhead of tribal nationalism, and the doctrine of the Chosen People presaged modern racism (1992, pp. 188–90; 1945, Vol. 1, pp. 6–8, Chap. 2, n. 3; Vol. 2, pp. 21–22). Rabbinic Judaism shut the Jews off from the world for two millennia, and the ghetto was the ultimate closed society, a "petrified form of Jewish tribalism" (1945, Vol. 2, Chap. 11, n. 56). Its inhabitants lived in misery, ignorance, and superstition, their separate existence evoking the suspicion and hatred of non-Jews and fueling antisemitism. It was imperative that the ghetto be opened and that enlightenment should follow. Integration was the only solution to the Jewish problem.

For Popper, Jewish nationalism reaffirmed tribal bonds, and was a colossal mistake. Until the last decade of his life, he rarely expressed himself in public about Zionism or the Jewish Question, but he poured out his wrath on Israel in private. His single major treatment of the Jewish Question is in his *Autobiography* (1976, Sect. 21). It proved highly controversial. He made no distinction between Diaspora nationalism and cultural or political Zionism: All were misguided. Reviving the ancient language (Hebrew) in Europe was no more legitimate than colonization in Palestine. As an ethnonational response to antisemitism, Jewish nationalism was bound to increase hatred of the Jews in Europe, and Zionism was sure to incite a new conflict with the Arabs in Palestine. Israel was a tragic error. "The status quo is the only possible policy in that maze of nations which peoples Europe and the Near East" (draft of 1976). Once Israel was founded, there was no way to undo the mistake, and he "strongly opposed all those who sympathize with the Arab attempts to expel the [Jews]" (Ibid.). But he remained highly critical of the Jewish state, insisting that its "racial" character give way to equal citizenship.

A viable Jewish diaspora could have solved Popper's quandary, but, in a world threatened by the National Socialists or dominated by their memory, a secure diaspora appeared just as much a dream as cosmopolitanism. A separate Jewish community, however acculturated, thought Popper, endangered the Jews. Antisemitism was to be feared in all places and times, and assimilation was the only viable response: "[A]ll people of Jewish origin [had] to do their best not to provoke [antisemitism]." The Jews did the opposite. They "insisted that they were proud of their [racial origins]" and triggered a racial war that brought their own destruction (Ibid.). They "invaded politics and journalism," drawing attention to their wealth and success, and, assuming leadership positions among the Socialists, they contributed to fascism's triumph (Ibid.). In his critique of Austrian Jews, Popper descended from cosmopolitanism to a position dangerously close to the antisemitism he feared. The increasing nationalization of Central European politics overwhelmed his cosmopolitanism. Jews were not to expect the fulfillment of cosmopolitanism but to accommodate themselves to ethnonationalism. They had to disappear as Jews. They could only become cosmopolitan citizens in a Kantian Kingdom of Ends.

The antinomies of cosmopolitanism meant that Popper did not leave much more room for the assimilated Jew than he did for the Zionist. Jews were citizens of the cosmopolitan empire but advised to keep a low profile in politics. He gave up on imperial pluralism's promise to the Jews to be recognized as a nationality, and yet he could not make good on his cosmopolitan empire, which turned out to be utopian. His difficulties were, however, historically contingent, due to fascism, and not essential to his cosmopolitanism. Contemporary Europeans negotiate more easily between pluralism and cosmopolitanism and recognize the Jews as European. Popper's cosmopolitanism can appeal to them in ways more immediate than Habsburg imperial pluralism. His democratic cosmopolitan empire provides a historical link between late imperial Austria and contemporary Europe. Presaging empire's rehabilitation in contemporary academic discourse, Popper made a progressive Austrian national narrative leading from the Monarchy through the First and Second Republic to a European Austria seem attractive (compare: Hardt and Negri 2000; Burbank and Cooper 2010). We have not yet heard the last of his cosmopolitanism.

This chapter reworks material from my *Jacob & Esau: Jewish European History Between Nation and Empire* (Cambridge UK: Cambridge University Press, forthcoming 2018).

References

Belke, Ingrid. 1978. *Die sozialreformerischen Ideen von Joseph Popper-Lynkeus (1838–1921).* In *Zusammenhang mit allgemeinen Reformbestrebungen des Wiener Bürgertums um die Jahrhundertwende.* Tübingen: Mohr.

Beller, Steven. 1989. *Vienna and the Jews: A Cultural History, 1867–1938*. New York: Cambridge University Press.

———. 1996. Patriotism and the National Identity of Habsburg Jewry, 1860–1914. *Leo Baeck Institute Yearbook* 41: 215–238.

Boyer, John. 1978. Freud, Marriage, and Late Viennese Liberalism: A Commentary from 1905. *Journal of Modern History* 50: 72–102.

———. 1995. *Culture and Political Crisis in Vienna: Christian Socialism in Power, 1897–1918*. Chicago: University of Chicago Press.

Brubaker, Roger. 1996. *Nationalism Reframed*. New York: Cambridge University Press.

Burbank, Jane, and Frederick Cooper. 2010. *Empires in World History*. Princeton: Princeton University Press.

Feigl, Herbert. 1969. The 'Wiener Kreis' in America. In *The Intellectual Migration: Europe and America, 1930–1960*, ed. Donald Fleming and Bernard Bailyn. Cambridge: Harvard University Press.

Fuchs, Albert. 1949. *Geistige Strömungen in Österreich, 1867–1918*. Vienna: Löcker.

Gellner, Ernst. 1998. *Language and Solitude: Wittgenstein, Malinowski and the Habsburg Dilemma*. Cambridge: Cambridge University Press.

Hacohen, Malachi H. 1996. Karl Popper in Exile: The Viennese Progressive Imagination and the Making of *The Open Society*. *Philosophy of the Social Sciences* 26: 452–492.

———. 2000. *Karl Popper – The Formative Years, 1902–1945: Politics and Philosophy in Interwar Vienna*. New York: Cambridge University Press.

———. 2002. La città celeste di Popper: Platone, Atene e la società aperta. In *Karl R. Popper, 1902–2002: ripensando il razionalismo critico*, ed. Stefano Gattei, vol. 2, 12–33. Bologna: Analisi-Trend.

———. 2006. The Young Popper as a Scholarly Field. In *Proceedings of the Karl Popper: A Centenary Assessment*, ed. Ian Jarvie, Karl Milford, and David Miller, vol. 1, 99–110. Aldershot, Hants, UK: Ashgate Publishers.

———. 2018. *Jacob & Esau: Jewish European History between Nation and Empire*. Cambridge: Cambridge University Press.

Hainisch, Michael. 1892. *Die Zukunft der Deutsch-Österreicher*. Vienna: F. Deuticke.

Hall, John A. 2010. *Ernest Gellner: An Intellectual Biography*. London: Verso.

Hardt, Michael, and Antonio Negri. 2000. *Empire*. Cambridge: Harvard University Press.

Hertzka, Theodor. 1890. *Freiland: Ein soziales Zukunftsbild*. Leipzig: Duncker & Humblot.

Holleis, Eva. 1978. *Die Sozialpolitische Partei: Sozialliberale Bestrebungen in Wien um 1900*. Munich: Oldenbourg.

Judson, Pieter. 2007. *Guardians of the Nation*. Cambridge: Harvard University Press.
King, Jeremy. 2002. *Budweisers into Germans and Czechs*. Princeton: Princeton University Press.
Marx, Karl. 1961. Zur Kritik der Hegelschen Rechtsphilosophie. Einleitung. *Werke*, 41 vols. Berlin: Dietz, 1: 390.
Menger, Anton. 1904. *Neue Staatslehre*. Jena: G. Fischer.
Popper, Karl R. 1927. Zur Philosophie des Heimatgedankens. *Die Quelle* 77: 899–908.
———. 1945. *The Open Society and Its Enemies*, 2 vols. London: Routledge & Sons.
———. 1959. *The Logic of Scientific Discovery*. London: Hutchinson. Originally: *Logik der Forschung: Zur Erkenntnistheorie der modernen Naturwissenschaft*. Vienna: Springer, 1935.
———. 1963. *Conjectures and Refutations*. New York: Basic Books.
———. 1976. *Unended Quest: An Intellectual Autobiography*. La Salle, IL: Open Court.
———. 1992. *In Search of a Better World*. London: Routledge.
———. 1994a. *The Myth of the Framework*. London: Routledge.
———. 1994b. Prague Lecture. www.lf3.cuni.cz/aff/p2_e.html (access through The Karl Popper Web: www.eeng.dcu.ie/~tkpw/).
———. n.d. Draft of Autobiography (1976). Hoover Institute Archives, Popper Papers (134, 4, 9, 12).
Popper-Lynkeus, Joseph. 1912. *Die allgemeine Nährpflicht als Lösung der sozialen Frage*. Dresden: Carl Reissner.
Rozenblit, Marsha. 1983. *The Jews of Vienna: Identity and Assimilation, 1867–1914*. Albany, NY: State University of New York Press.
Schorske, Carl. 1980. *Fin-de-Siècle Vienna: Politics and Culture*. New York: Vintage.
Simkin, Colin. 1993. The Birth of *The Open Society*. In *Popper's Views on Natural and Social Science*, ed. Colin Simkin, 183–190. Leiden, NY: Brill.
Sorrels, Katherine. 2016. *Cosmopolitan Outsiders: Imperial Inclusion, National Exclusion, and the Pan-European Idea, 1900–1930*. New York: Palgrave Macmillan.
Stadler, Friedrich. 1981. Spätaufklärung und Sozialdemokratie in Wien, 1918–1938. In *Aufbruch und Untergang: österreichische Kultur zwischen 1918 und 1938*, ed. Franz Kadrnoska, 441–473. Vienna: Europaverlag.
———. 1997. *Studien zum Wiener Kreis*. Frankfurt am Main: Suhrkamp.
Wittelshöfer, Otto. 1894. Politische und wirtschaftliche Gesichtspunkte in der österreichischen Nationalitätenfrage. *Preußische Jahrbücher* 76:3. Berlin: Walther.

Wittgenstein, Ludwig. 1922. *Tractatus Logico-Philosophicus.* Trans. C.K. Ogden. New York: Harcourt, Brace & Co.

Zimmern, Alfred. 1911. *The Greek Commonwealth: Politics and Economics in Fifth Century Athens.* London: Oxford University Press.

———. 1934. *The Third British Empire.* London: Oxford University Press.

CHAPTER 17

Popper and Hume: Two Great Skeptics

Zuzana Parusniková

Introduction

Popper explicitly discusses two problems in Hume's epistemology. He praises Hume for his critique of induction, specifically for his claim that inductive inferences are logically invalid. He rejects Hume's psychological account of induction, specifically his theory of belief formation by repetition. Thus, Popper famously concludes, Hume buried the logical gems in the psychological mud, and became a believer in an irrationalist epistemology. The logical problem of induction gives Popper the impetus for spelling out his new, negative concept of reason, one which is incompatible with justification. However, Popper's approach does not adequately deal with all the relevant themes related to Hume's psychological problem of induction, mainly with the theme of our instinctive yearning for justification. Yet, Popper and Hume have more in common than Popper explicitly acknowledges. On the broadest level they both rebel against the established paradigms of rationality. Popper denounces all traditional rationalism in which reason provides good reasons for knowledge claims. For him, reason can only destabilize these claims, but through finding errors we discover new problems to be solved and thus make knowledge grow.

Z. Parusniková (✉)
Institute of Philosophy, The Czech Academy of Sciences,
Prague, Czech Republic
e-mail: parusnikova@volny.cz

R. Sassower, N. Laor (eds.), *The Impact of Critical Rationalism*,
https://doi.org/10.1007/978-3-319-90826-7_17

Hume takes a different route showing the impotence of reason and dismissing its foundational role for science. Under the influence of ancient skepticism he advises us to accept this predicament and thus attain peace of mind. Hume, however, viewed skepticism not only as an epistemological exercise but as an approach to life. Though he retained radical skepticism toward knowledge, he implemented mitigated skepticism in common life, morality, and society. Both Popper and Hume believed that it was through the cultivation of our cognitive and moral faculties that mankind would progress. In this sense, both Hume and Popper show their Enlightenment affiliation.

Hume in Vienna

In the first half of the twentieth century, Vienna became an important center of modern analytical philosophy. At the turn of the century, the leading figure in the positivist movement had been Ernst Mach, followed in the interwar period by Moritz Schlick and the group of philosophers, logicians, and scientists, named the *Wiener Kreis*, or the *Vienna Circle*. David Hume played a prominent role in these philosophical developments, as positivists followed two themes in his epistemology. First, they were determined to *eliminate metaphysics* from philosophy, and thus introduce philosophy as a scientific discipline providing the rational foundations of natural science; Hume targeted scholasticism and the Church dogmas, while the Viennese philosophers stood mainly against German idealism. Second, they adopted the *empiricist* position and exploited Hume's conception of sense perception as the ultimate source of all knowledge.

Inspired by Ludwig Wittgenstein, the Vienna Circle developed the method of the logical analysis of language by which only those statements which can be empirically verified are scientific: language must be purged of all metaphysical elements. The immediate experience of the senses is verified by its sense-givenness and expressed in the simplest, atomic statements ("protocol sentences"). More complex statements are built according to the logical rules of syntax, as suggested by the title of Rudolf Carnap's famous book, *Der logische Aufbau der Welt* (1928), and these statements must be reducible to the atomic statements from which they are made. This procedure loosely draws on Hume's empiricist principle that "all our ideas or more feeble perceptions are copies of our impressions or more lively ones" (2005a, p. 19); general statements are derived from impressions and can be traced back to them. Logical positivists went far beyond Hume in their claim that verification entails not only the truth but also the meaning of a statement; all other statements are considered metaphysical and therefore meaningless, as Friedrich Waismann summarizes in the famous phrase: "[I]f there is no possible way to determine whether a statement is true then that statement has no meaning whatsoever. For the meaning of a statement is the method of its verification" (1930, p. 229).

Positivists were all in agreement with Hume that

> [I]t is a question of fact, whether the perceptions of the senses be produced by external objects, resembling them: How shall this question be determined? By experience surely; as all other questions of a like nature. But here experience is, and must be entirely silent. The mind has never any thing [sic.] present to it but the perceptions, and cannot possibly reach any experience of their connexion with objects. (2005a, p. 153)

Philosophy is about analyzing the *neutral stream of sensations* and prescribing the rules regarding how they should be assembled—in Hume's terminology, how the bundles of perceptions are formed from those that make their first appearance in the soul and strike us with great force and vivacity. Positivists developed Hume's distinction between the relations of ideas (analytic a priori; logic, mathematics) and matters of fact (synthetic a posteriori; propositions with empirical content) and committed to the flames everything beyond that (synthetic a priori); such propositions belong to metaphysics and are examples of a speculative, obscure, and abstruse gibberish.

However, we must not forget another Viennese philosopher of that time, Karl Popper. Popper absorbed the influences of logical positivism, especially the concern for science and the respect for logic, yet criticized positivists on most philosophical issues; Otto Neurath labeled him the "official opposition." Yet Popper, Hume, and the logical positivists shared one general aim: to formulate a *new philosophy* linked to new scientific discoveries. Hume and the Viennese group of philosophers (and the *Berliner Gruppe*) witnessed respectively the emergence of a new scientific paradigm. In Hume's time, the most celebrated figure of the age was, of course, Isaac Newton, the discoverer of the laws and forces which govern nature and the universe; for Hume, he was "the greatest and rarest genius that ever arose for the ornament and instruction of the species" (2005b, p. 374). The positivists and Popper lived and worked in the time of Albert Einstein's discoveries and were engaged in both scientific and philosophical communication with him; Einstein was influenced especially by Mach both in physics (Mach's principle) and in philosophy (concepts must be grounded in experience; experimental science is based on observation and measurement). (The extensive literature concerning this context of Einstein's views cannot be discussed here; see Holton 1973; Howard 2004; Norton 1993, 2009.) Einstein realized the skeptical implications of Hume's empiricism (and of Mach's endorsement of Hume). In his autobiographical notes, he says that Mach's ideal of science, consisting in a mere description and ordering of experience, lacks the creative element needed for scientific discoveries. What he drew from the Humean legacy was the definition of concepts (Hume's "ideas"): they must be grounded in experience, or they are fictional; this impacted directly on his science. In his words:

> [Concepts] must be grounded in experience and, in so far as they extended beyond that grounding, they are fictional or arbitrary. Rather than merely abjuring these fictional concepts (Mach) or at best tolerating them (Hume), Einstein saw them as revealing an arbitrariness in our physical theorizing. The fictional concept could still be introduced through a freely chosen definition, as long as the definition did not tacitly commit us to false presumptions. That view of concepts enabled Einstein to abandon the notion of absolute simultaneity when he finally realized that this notion was all that obstructed his conforming of Maxwell's electrodynamics to the principle of relativity. He replaced it by a new notion of simultaneity introduced through a definition that did not commit him to the absoluteness of simultaneity. (Norton 2009, pp. 379–383)

All these philosophers believed that new scientific paradigms open a new era in the development of mankind, and that philosophy must play an active role in this process by drawing as close to science as possible. Philosophy should *serve* science by providing strict methodological rules to produce a scientific conception of the world. Hume was interested in the science of man, but he, too, viewed its first part, epistemology, as the basis for all other sciences; his declaration fits in with the Enlightenment optimistic spirit: "[I]n pretending therefore to explain the principles of human nature, we in effect propose a compleat [sic.] system of the sciences, built on a foundation almost entirely new, and the only one upon which they can stand with any security" (1981, p. xvi); the fact that Hume failed to fulfill this task is a different matter. The positivists and Popper endorsed this aim.

On fundamental philosophical principles, Popper stands in opposition to positivism on most counts. He did not eliminate metaphysics from meaningful discourse, but showed that "good" metaphysics (e.g., atomism) can be an inspiration for science. He endorsed realism which may be a metaphysical assumption, but one that provides the best framework for the development of science; therefore, it is accepted as an act of decision. He saw the scientific enterprise as more adventurous than the positivist arduous logical construction of sentences; according to Popper, this consists in *bold conjectures* which are *empirically tested* with the aim of *refuting* them, not of verifying, confirming, or justifying them in any way. He put great emphasis on the dynamic, creative, and processional aspects of science. He denied that pure observation is possible, since it is always theory- and problem-laden.

Accordingly, Popper focused on entirely different issues in Hume's philosophy. Unlike the positivists, he accepted the *skeptical aspects of Hume's epistemology*. He admitted that the logical problem of induction was irresolvable, but denied that this conclusion should endanger the rational status of science. Against both Hume and logical positivism, he argued that it is induction itself that needs to be reconsidered as the method of science, and that the justification of general statements should no longer define the standards of rationality. Popper's position thus unfolds from his critique of induction and, ultimately, from his confrontation with Hume.

Induction Is a Myth: Hume's Logical Gem

Popper fully accepts Hume's skepticism concerning induction; he opens his *Logik der Forschung* (1934) with this theme and later elaborates on it mainly in *Objective Knowledge* (1972). The problem of induction arises from the rule of empiricism, namely, that "we have no reason to draw any inference concerning any object beyond those of which have had experience" (Hume 1981, p. 139); yet Hume adds: "[A]ll our reasonings from experience [of which causation is the case in point] are founded on the supposition, that the course of nature will continue uniformly the same" (Ibid., p. 651). Here, the skeptical dilemma fully displays itself: our natural inductive reasoning concerning matters of fact transcends experience and thus lacks rational foundations. It is based solely on custom: the expectations that the future will be like the past arise due to the *associative mechanism* of our minds. After repeated experience of two things conjoined together, we become "accustomed to the uniformity of nature, we acquire a general habit, by which we always transfer the known to the unknown, and conceive the latter to resemble the former" (Ibid., p. 107, n. 1).

In order to legitimize inductive inferences, we have to establish the validity of the principle of induction or, in Hume's words, the Uniformity Principle. This could be done either by proof (demonstration) or by probability (reasoning from experience). The former is impossible since proof requires certainty, while the Uniformity Principle is contingent; "'tis possible that the course of nature may change, since we can conceive such a change" (Ibid., p. 651). Neither can experience provide proof, because it, too, is predicated upon the uniformity of nature. As Hume remarks: "[A] ll probable arguments are built on the supposition, that there is this conformity betwixt the future and the past … this conformity is a *matter of fact*, and … will admit no proof but from experience. But our experience in the past can be a proof of nothing in the future" (Ibid., pp. 651–652). We are in a closed circle; the only two choices left are infinite regress or the dogmatic acceptance of the Uniformity Principle.

Popper endorses Hume's arguments and claims that as for an empirical proof of the validity of inductive inferences, "to justify these we would have to assume an inductive principle of a higher order; and so on. Thus the attempt to base the principle of induction on experience breaks down, since it must lead to infinite regress" (Popper 1965, p. 29). This verdict is also acknowledged by Bertrand Russell; "[The] principle itself cannot, of course, without circularity, be inferred from observed uniformities, since it is required to justify any such inference" (1946, p. 699). Russell tries to avoid this deadlock of proving the principle of induction as follows: "we shall nevertheless hold some knowledge is *a priori* in the sense that experience which makes us think of it does not suffice to prove it, but merely so directs our attention that we see its truth without requiring any proof from experience" (1912, p. 116). After the failure to prove the principle of induction, either as an analytic a priori or a synthetic a posteriori, only the possibility of its being the synthetic a priori is left, and this possibility must be dismissed as metaphysical. This deadlock concerning the rational foundation of inductive reasoning has a long history dating to the ancient skeptics; an example is the second and the fifth skeptical modes (of the five ascribed to Agrippa). The second mode posits the infinite regress, and the fifth reciprocal mode exposes that "what ought to be confirmatory of the objects under investigation needs to be made convincing by the object under investigation" (Annas and Barnes 2000, p. 41; see also pp. 166 and 169).

There have been many attempts to solve the problem of induction through an appeal to probability. In Hume's time, this strategy was applied by the French skeptics, Pierre Gassendi and his circle (who influenced John Locke), and the next generation was represented primarily by Pierre Daniel Huet. Being scientists themselves, they were concerned with the foundations of science and pragmatically considered Academic (probabilistic) skepticism to be both *suitable* for *reducing skeptical doubts* and *sufficient* for *establishing the rational status of science*. They claimed that the disposition of falsity arises not from perceptions but from generalizations extrapolated from those perceptions, and recommended caution and modesty. Hume read the then very popular Huet's *Traité philosophique de la foiblesse de l'esprit humain* (1723), which had been translated into English by 1725. Huet's aim was to show that "Truth cannot be perfectly and with certainty known to human Understanding" but that "in all Things we ought to follow Probability" (Huet 1725, p. 86). Hume saw that this solution was logically inconsistent and insisted that inductive inference added nothing to the probability of a conclusion. On this point, Hume follows the radical skeptical tradition developed by Michel de Montaigne and Pierre Bayle; Montaigne asks, "how can they [the Academics] bring themselves to yield to verisimilitude if they cannot recognize verity? How can they know there to be a resemblance to something the essence of which they do not know?" and answers, along with Hume, that "[W]e judge entirely, or entirely not" (1993, p. 141). Popper, too, was surrounded by probabilistic solutions aiming to legitimize inductive inferences. Many logical positivists (e.g., Hans Reichenbach and Carnap) argued that it is not possible to attain either truth or falsity in knowledge; rather, scientific statements can only attain continuous degrees of probability, the unattainable upper limit of which is truth. Popper rejects this solution and says that nothing is gained if the principle of induction is taken as probable, "[F]or if a certain degree of probability is to be assigned to statements based on the inductive inference, then this will have to be justified by invoking a new principle of induction" and so on ad infinitum (1965, p. 30). Popper declared that induction is a myth, adding, with his typical audacity, that it is he himself who must admit responsibility for the death of logical positivism (1982, p. 88). He saw Hume as his ally in exposing the logical incoherence of inductive inferences.

In contemporary Hume scholarship, this skeptical interpretation of Hume is considered old-fashioned (or more kindly, just "traditional"). Various rescue strategies for Humean inductive science are considered, for instance, appealing to Hume's *naturalism* in which our natural inductive inclinations are self-justified and hence custom and belief provide a sufficient basis for empirical [inductive] science (Millican 2006). Alternatively, Hume's contribution is seen predominantly as "mental geography" consisting in what we call today "cognitive science" (without the skeptical flavor); among others, Don Garrett claims that reasoning itself is a natural activity and reason *already arises from our natural needs*, best suited to satisfy those desires we naturally have (1997, 2004). Tom Beauchamp and Thomas Mappes (1975) even deny there is a problem of induction in Hume's philosophy; Hume apparently ties this problem to the normative conception of reason, a conception which he rejects. Some (e.g., John Searle and David Papineau) have dismissed the problem as a mere puzzle which is "a good issue for first-year philosophy students to cut their teeth on. But outside the classroom nobody seriously doubts that we do know about tables" (Papineau 1995, pp. 4f). Also, some attempts have been made, especially by the *New Humeans*, to introduce Hume as an ontological realist (Read and Richman 2007) in order to remove the skeptical element of phenomenalism.

This trend in Hume scholarship is based on the assumption that finding some science-friendly interpretation of Hume—and thus weakening his skepticism—would "rehabilitate" Hume and somehow increase his philosophical value. This approach is symptomatic of the dominant analytical interpretations which share the assumption that philosophy must comply with science. I claim that naturalistic or realist interpretations fail to appreciate the historical context of Hume's philosophy. Hume lived in a different time, when (a) epistemological questions were of primary importance, (b) epistemology was considered a normative discipline, and (c) a rational grounding of knowledge was required. Besides, it is precisely the consistent skepticism regarding induction (and causation) that makes Hume's contribution so interesting and influential on further developments in epistemology. This influence is particularly noticeable in the case of Popper; Hume's skepticism regarding induction inspired him to develop his original hypothetico-deductive model of understanding.

Induction Is a Myth: Hume's Psychological Mud

Popper criticizes Hume for shifting the logical problem of induction into the psychological field. Though we cannot rationally justify the view that nature will continue in a uniform manner, explains Hume, we *believe* it.

Belief is the result of the repeated experience of the constant conjunction of two objects, and Hume's example is billiard balls. "[M]y mind runs by habit from the visible object of one ball moving towards another, to the usual effect of motion in the second ball. It not only conceives that motion, but *feels* something different in the conception of it from a mere reverie of the imagination" (1981, pp. 653–654).

Popper criticizes Hume for not having gone far enough, for not having considered that *induction plays no role either in belief formation or knowledge acquisition*. Thus, Popper concludes that Hume "buried the logical gems in the psychological mud" (1979, p. 89). Popper claims that "induction—the formation of a belief by repetition—is a myth" (Ibid., p. 23) and argues that the expectation of regularities is inborn, genetically given. He sides with Immanuel Kant but with a proviso: genetic a priori dispositions may change in the course of evolution. By this step, Popper reduces the psychological problem of induction to a biological one: "psychology should be regarded as a biological discipline" (Ibid., p. 24). Popper developed this theme in his influential conception of evolutionary epistemology. This leads to another aspect of Popper's critique of Hume's psychological problem of induction. Contrary to the beliefs of Hume and the positivists, there is no *pure* observation. Every observation is already *theory-laden*, determined by challenging theoretical problems in science and by pressing practical problems in life. We approach both the world of theories and the ordinary world with some expectations which, if not fulfilled, give rise to *problems* we need to solve, either to enrich knowledge or merely to survive. Therefore, Popper notes that induction "is neither a psychological fact, nor a fact of ordinary life, nor one of scientific procedure" (1985, p. 53).

Furthermore, as in the evolution of life, also in science only the *fittest* (hypothesis) survives. As noted above, Popper's epistemology is based on the method of *trial and error elimination* in which bold conjectures undergo a confrontation with reality via tests that aim at disproving the empirically testable predictions of the hypothesis; this phase guarantees the *objectivity* of knowledge. The discovery of errors opens up new fields of enquiry and thus drives knowledge forward. Popper lets criticism hit with full blast since we can let our theories die (1979, p. 78). This implies a positive attitude to error; since we do not fear for life, we can actively pursue criticism and embrace its stimulating effects. Hume, however, did not give up the inductive model of learning and thus faced a clash with its rational indefensibility. This led him to give up the rational status of knowledge and to transfer knowledge to the domain of natural instincts. As Popper puts it, Hume ended up as a "believer in an irrationalist epistemology" (Ibid., p. 4). Popper's negative assessment of Hume seems too strong, and expresses Popper's rationalistic stance that for Hume, our natural, instinctive faculties, though not legitimized by rational reflection, are sustained by the wisdom of nature.

Popper thus rejects the psychologism entailed in the "givenness" of sense perceptions (mind as a bucket) and the justification of the subsequent generalizations. Popper's solution to the psychological problem of induction seems clear-cut: he presents a consistent deductive methodology based on criticism and no justificationism. However, there is one problem in his solution that undermines his model: whether our expectations of regularities arise due to repeated experience or are inborn, the fact is that *we have them* irrespective of their origins. Moreover, our expectations contain a strong need for confirmation, encouraging us to justify rather than falsify our beliefs. We simply *believe* that the greater the positive evidence, the more support (the higher the probability) there is for the general statement. Even if our expectations of regularities are inborn, we still *feel* an irresistible desire that they be fulfilled. Yet for Popper, any empirical "support" of a theory is illusory; only the falsification attempts are rational, and only falsifying evidence is deemed relevant. Popper's methodological norms do not weaken our justificationist inclinations that "sabotage" the power of criticism, and thus the authority of reason. Ultimately, Popper's epistemology seems to suffer from a similar clash between the rational and the natural as does Hume's.

The Skeptical Challenge

Hume's skepticism concerning induction unfolds from the requirement that unless the inductive principle is valid, science does not have a rational warrant. Even the positivist rescue move, inferring a greater or a lesser probability of truth from given premises, presupposes the validity of the principle of induction; and since it cannot be proved empirically (infinite regress), we have to accept it a priori which, in turn, defies empiricism. Popper chooses a different path. Instead of searching for some justification of induction, he offers *a new paradigm of rationality free from induction and justification*. We have to dispense with the principle of induction and seriously consider the possibility "that we can, and must, do without induction" (Ibid., p. 28); after all, induction is a myth both in the logical and the psychological sense. His elimination of induction has consequences for other qualities expected from a rational epistemology.

Justification, certainty, credibility, reliability, and probability of scientific theories are for Popper just crutches in our search for an illusory safety. We have been fed for centuries the justificationist doctrine and become accustomed to the idea that statements must be justified in order to "have weight." Popper, however, introduces a new criterion of demarcation consisting of the empirical *falsifiability* of theories; he emphasizes that falsification is not just the other side of the verificationist coin but that there is an asymmetry between them as a result of the logical form of universal statements: a theory can be falsified by a single observational statement. Popper also rejects the charge that singular observational statements cannot be proved true, since they contain universals, and therefore cannot be used for the decisive falsification of a theory. Popper never claimed that basic statements can be proved true: we test them by observation until we decide to accept them (basic statements are accepted as a result of a decision or agreement). In the process of bold conjectures and their attempted refutations, falsehoods are eliminated from knowledge, and those theories which are not falsified are kept in science, merely *kept*, not *supported* by the fact that their falsification failed: any support would already be an inductive move.

Critical rationalism boosts the growth of knowledge. Justification—besides being logically incorrect—is a defensive attitude that prevents us from daring (improbable) guesses, suppresses intellectual conflicts, and fails to exploit the creative potential of erring. However, if we take falsification as the desirable goal of testing, we are not afraid of hazarding bold (improbable) guesses in the hope of exploiting our errors as a source of new, deeper *problems*. Criticism requires an adventurous spirit and expresses the high-risk nature of the cognitive process. Justification and content are inversely related, and therefore the aim of achieving the highest possible probability of our statements inhibits our willingness to take risks, and stifles the boldness and creativity of our thinking. Popper's orientation toward the negative powers of reason marked an *epistemological revolution* which redefined the methods of science and the nature of rationality. (Popper's conception of rationality has always provoked argument. He has often been accused of being too negative, too extreme; his falsificationism has been said to be applicable only to single theories and thus unsuitable for science which operates within broader scientific programs or paradigms. I cannot address these issues within the scope of this paper.)

It takes some courage to change the perspective and to view rationality in negative—yet optimistic—terms (Miller 1994; Parusniková 2009). In Popper's account, the rational method allows only expulsion procedures but no confirming proofs of theories. Therefore, his conception of *critical rationalism* cannot be described merely as a promotion of criticism, but as a new *negative model of reason* constituted by the imperative of falsification. Popper's model of reason accepts uncertainty as a predicament which does not endanger truth as the goal of knowledge. Experience "provides no positive support for [theories], provides no inductive lift, leaves them floating in the ocean of uncertainty" (Watkins 1984, pp. 353–354). Living in uncertainty, however, is not easy to accept. As suggested above, Popper's solution to Hume's psychological problem of induction underestimates our natural need for reassurance, stability, and safety, which is reflected in seeking positive evidence that we *believe* confirms our expectations, but without a rational warrant. In other words, living in permanent uncertainty is traumatic. Popper tries to overcome this problem by introducing the *principle of transference*; this principle implies that what is true in logic is true in psychology. Popper restates Hume's psychological problem of induction as follows: "If we look at a theory critically, from the point of view of sufficient evidence rather than from any pragmatic point of view, do we always have the feeling of complete assurance or certainty of its truth, even with respect to the best-tested theories, such as that the sun rises every day? I think that the answer here is: No" (1979, p. 26).

Popper classifies the belief that the sun rises every day as a *pragmatic* belief. However, this does not prevent him from repeating that after we have rationally considered the available evidence, "we shall have to admit that the sun may not rise tomorrow over London after all" (Ibid.). There is a crucial difference between *admitting*, after a rational consideration, and *feeling* compelled by inborn instincts. The principle of transference, although removed from normative scientific methodology (based on falsification), does not solve the Humean dilemma which counterposes reason to our natural and dogmatic inclinations. By the principle of transference, Popper hopes to strengthen the position of reason at the cost of instincts; this expresses his endorsement of the modern ideal of the maturing of mankind, accompanied by the advance of reason. In the evolutionary process, human minds will increasingly internalize the rational (critical) attitude which will, in due course, overpower instincts, possibly causing a change in our genes. This reflects, as Hacohen remarks (2000, p. 69), the typical progressive intellectual atmosphere in Vienna—termed *Spätaufklärung*—that stood in opposition to various conservative, clerical, and nationalist tendencies, and felt the need to reaffirm the scientific ethos of the Enlightenment. However, the principle of transference seems insufficient for tempering our justificationist instincts; this casts doubts on Popper's solution of the psychological problem of induction.

Hume, by contrast, based his epistemology on inductive cognitive processes and our tendency to justify our expectations. As a result, he consistently held a radical skeptical view concerning the rational foundations of knowledge and shifted the problem of knowledge to the area of natural instincts as the ultimate authority: "I must yield to the current of nature, in submitting to my senses and understanding ... If I must be a fool, as all those who reason or believe any thing [*sic*] *certainly* are, my follies shall at least be natural and agreeable" (1981, pp. 269–270). The *nonchalance* with which Hume brushed away the prominent epistemological concerns of the day, the fact that he shrugged them off with a mere pronouncement, "if we believe, that fire warms, or water refreshes, it is only because it costs us too much pains to think otherwise" (Ibid., p. 270), is astonishing, even *scandalous* for a modern philosopher.

For Popper, this was an unnecessary resignation. Hume himself did not consider this (skeptical) epistemological predicament a resignation, and accepted the weakness of our reason with truly skeptical calm. Epistemology was only one part of the new *science of man* comprising, besides epistemology, of common life, morality, politics, and society in general. In these areas, he was an optimist due to his full appreciation of ancient skepticism which is not only an epistemological exercise, but *a way of life*. Philosophy, then, must teach us to *accept the wretched state of reason* and thus stop torturing our minds with insoluble contradictions. Therefore, Hume decided to *leave the epistemological problem unsolved and to leave us perplexed*. In a truly Pyrrhonian fashion, he accepted that the impossibility of grounding knowledge in reason is an expression of the whimsical condition of mankind. Instead of trying to overcome epistemological skepticism in any theoretical arguments, he decided to *relax this (skeptical) bent of mind* in order to achieve mental balance.

The original Pyrrhonian connotations of *ataraxia* were tied to the quietude of mind and to detachment from life. Hume changed this ideal by placing it within the bustle of life. While the Pyrrhonists put an emphasis on the contemplative character of tranquility, Hume added a more frivolous flavor and emphasized the elements of pleasure and joy. Carefree acceptance of the limitations of reason, taking one's mind off skeptical pessimism by being merry with friends, or walking along the river and absorbing the beauty of nature, getting involved in trade and business, discussions, and politics—these are Hume's recipe for a *happy life*. Though he retained radical skepticism toward knowledge, he implemented mitigated skepticism in the areas of common and social life.

In that context, Hume draws on Cicero's emphasis on moderation and on linking philosophy to society and to its citizens' involvement in public matters. The lesson taken from Pyrrhonism, besides epistemological skepticism, consists in our being aware of our limitations and applying modesty and caution to all our attitudes and actions. The lesson taken from *Academica* is "to apply their [philosophers'] own practical wisdom and insight to the service of humanity … Hence it follows that the claims of human society and the bonds that unite men together take precedence of the pursuit of speculative knowledge" (Cicero 1913, pp. 156–157). This resonates with Hume's conviction that critical, moderately skeptical, and socially sensitive reason is the only faculty that can instruct us in "the tendency of qualities and actions, and point out their beneficial consequences to society and to their possessor" (2005a, p. 285).

Therefore, true philosophy avoids "all distant and high enquiries, confines itself to common life, and to such subjects as fall under daily practice and experience" (Ibid., p. 161). Morality and politics can thus be set free of the pretentions of the doctrinaires and of the vanity of the dogmatists, fostering the values of modesty, tolerance, and mutual respect instead. In the area of social and political philosophy, Hume and Popper hold hands. Hume endorses the Enlightenment *ethos* of establishing a better social order, without the French arrogance of reason. The French model demands a radical break with the past: new foundations should be built on the debris of the old order. Accordingly, the religious tradition was replaced by the cult of reason that became as dogmatic and oppressive as the practices of the Church. Later, during the French revolution, in its most extreme form, desecrated churches served as temples of reason; each year a *Fête de la Raison* took place instead of Christian festivals. This *fête* had a grandiose form, often with the goddess of reason—in an outfit resembling the Greek Athena with a helmet—that was lowered down on a pedestal from the ceiling.

Both Hume and Popper promoted a different sort of social progress: critical, tolerant, and cautious. The revolutionary version of the Enlightenment contained all that Hume fought against: fanaticism and the lack of respect for tradition. For Hume, fanaticism was represented by religious enthusiasm, typical of "rash arrogance [and] lofty pretentions" (Ibid., p. 41). Hume favors a strategy that avoids excesses and is permanently concerned with providing the right circumstances for the development of an enlightened society that promotes mild sentiments and the dignity of virtue, ensuring that the welfare of humanity is not achieved at the cost of its members. Hume's vision is close to the Popperian ideal of the open society; both were fierce critics of institutions that oppress freedom of thought and forbid open discussion.

For Popper, the application of criticism and rationalism within society is the best defense against totalitarianism and utopianism. Utopian social engineering is built on the conviction that somebody—a leader, a philosopher, a class—has knowledge of those laws which should govern the future development of a society. This method, Popper says, "must lead to a dangerous dogmatic attachment to a blueprint for which countless sacrifices have been made" (1997, p. 163). Popper recommends that social change should be made through small steps which can be revised and reversed relatively painlessly. He calls this method "piecemeal social engineering" (or with more panache, "democratic social reconstruction" (1997, p. xviii)) which tackles specific problems rather than the ills of society as such. Of course, living in different times, Hume and Popper considered different forms of government which might meet these requirements. However, both rejected the possibility of quick, easy solutions to all social problems, the existence of an ideal society, and the euphoric strategies of sweeping revolutionary changes.

References

Annas, Julia, and Jonathan Barnes. 2000. *Sextus Empiricus: Outlines of Scepticism*. Cambridge: Cambridge University Press.

Beauchamp, Tom L., and Thomas Mappes. 1975. Is Hume Really a Sceptic About Induction? *American Philosophical Quarterly* 2 (12): 119–129.

Cicero. 1913. *De oficiis* [44 BC]. Trans. W. Miller. Cambridge: Harvard University Press.

Garrett, Don. 1997. *Cognition and Commitment in Hume's Philosophy*. Oxford: Oxford University Press.

———. 2004. A Small Tincture of Pyrrhonism: Skepticism and Naturalism in Hume's Science of Man. In *Pyrrhonian Skepticism*, ed. W. Sinnott-Armstrong, 68–99. New York: Oxford University Press.

Hacohen, Malachi H. 2000. *Karl Popper. The Formative Years, 1902–1945*. Cambridge: Cambridge University Press.

Holton, Gerald. 1973. Einstein, Mach, and the Search for Reality. In *Thematic Origins of Scientific Thought: Kepler to Einstein*, ed. Gerald Holton, 219–259. Cambridge: Harvard University Press.
Howard, Don A. 2004. Einstein's Philosophy of Science. *The Stanford Encyclopedia of Philosophy*, Summer 2010 Edition.
Huet, Pierre D. 1725 [1723]. *A Philosophical Treatise Concerning the Weakness of Human Understanding*. London: Gysbert Dommer.
Hume, David. 1981 [1739]. *A Treatise of Human Nature*. Ed. L.A. Selby-Bigge. Oxford: Clarendon Press.
———. 2005a [1748]. *An Enquiry Concerning Human Understanding*. Ed. L.A. Selby-Bigge. Oxford: Clarendon Press.
———. 2005b [1751]. *An Enquiry Concerning the Principles of Morals*. Ed. L.A. Selby-Bigge. Oxford: Clarendon Press.
Miller, David. 1994. *Critical Rationalism: A Restatement and Defence*. Chicago and La Salle, IL: Open Court.
Millican, Peter. 2006. The Context, Aims and Structure of Hume's First Enquiry. In *Reading Hume on Human Understanding*, 27–67. Oxford: Oxford University Press.
Montaigne, Michel. 1993 [1580]. *An Apology for Raymond Sebond*. Trans. M.A. Screech. London: Penguin.
Norton, John D. 1993. Mach's Principle Before Einstein. In *Mach's Principle: From Newton's Bucket to Quantum Gravity*, ed. J. Barbour and H. Pfister, 9–57. Boston: Birkhäuser.
———. 2009. How Hume and Mach Helped Einstein Find Special Relativity. In *Discourse on a New Method: Reinvigorating the Marriage of History and Philosophy of Science*, ed. M. Friedman, M. Domski, and M. Dickson, 359–386. La Salle, IL: Open Court.
Papineau, David A. 1995. Open Society, Closed Thinker. *The Times Literary Supplement* 23: 4–5.
Parusniková, Zuzana. 2009. *Ratio negativa* – The Popperian Challenge. In *Rethinking Popper*, ed. Zuzana Parusniková and Robert S. Cohen, 31–47. Netherland: Springer.
Popper, Karl R. 1965 [1934]. *The Logic of Scientific Discovery*. New York: Harper and Row.
———. 1979 [1972]. *Objective Knowledge. An Evolutionary Approach*. Oxford: Clarendon Press.
———. 1982 [1974]. *Unended Quest. An Intellectual Autobiography*. Glasgow: Fontana.
———. 1985 [1963]. *Conjectures and Refutations. The Growth of Scientific Knowledge*. London and Henley: Routledge & Kegan Paul.
———. 1997 [1945]. *The Open Society and Its Enemies*, Vol. 1. London: Routledge & Kegan Paul.

Read, Rupert, and Kenneth A. Richman, eds. 2007. *The New Hume Debate*. London and New York: Routledge.
Russell, Bertrand A.W. 1912. *The Problems of Philosophy*. New York: Henry Holt & Co.
———. 2004 [1946]. *History of Western Philosophy and Its Connection with Political and Social Circumstances from the Earliest Times to the Present Day*. Abingdon: Routledge.
Waismann, Friedrich. 1930. Logische Analyse der Wahrscheinlichkeitsbegriff. *Erkenntnis* 1: 228–248.
Watkins, John. 1984. *Science and Scepticism*. Princeton: Princeton University Press.

CHAPTER 18

The Tyranny in Science: The Case of Hugh Everett's Universal Wave Theory Formulation of Quantum Mechanics

Sheldon Richmond

Setting the Stage: Where Everett Enters the Debate

What was the stage like when Hugh Everett made his entry into the drama of quantum mechanics in the 1950s?

Before exploring the political situation of the Republic of Science in the early years of the Cold War and in the early years of the development of Big Science, a few words on Karl Popper and Ian C. Jarvie's notion of the Republic of Science. Jarvie (2001) argues that according to Popper's idealized model, methodological rules function as *social conventions*; and those social conventions are designed to function in opening up scientific knowledge to universal critical debate. Science is an institution or social technology whose aim is both to lessen error and incrementally approach and approximate the truth. Popper recognizes but underplays the external social and political pressures placed on science that can be somewhat corrected and minimized by exposing science as a social technology to democratic control. Furthermore, every institution requires social checks for those who have positions of power and authority. Jarvie does not dwell on

S. Richmond (✉)
Independent Scholar, Thornhill, ON, Canada

R. Sassower, N. Laor (eds.), *The Impact of Critical Rationalism*,
https://doi.org/10.1007/978-3-319-90826-7_18

the conservative role of tradition in science as a critical interpretation of the early Popper, where Popper (1945, 1957, 1959) underplays the power of tradition. Popper has a highly optimistic theory of tradition where institutions, instead of being dominated by tradition, can be open to the rational critique of tradition (1963, pp. 120–135). Though science as an institution can have a critical approach to tradition, the dominant function of tradition in social institutions, including science, is to treat traditional ideas as more or less immovable foundations (see Agassi and Meidan 2016, pp. 114 ff.; Jarvie 2001, pp. 212–231).

With the encouragement of his mentor John Archibald Wheeler, Everett visited Niels Bohr and corresponded with Bohr's associates. Everett found that his ideas as developed in the early stages of his dissertation were rebuffed. This, I think, was not mainly for intellectual reasons—though some were given—but for political reasons. For instance, given the logic of the situation where the Copenhagen school of quantum mechanics (from now QM) had basically established itself as forming a micro-tradition of intellectual research whose main intellectual leaders had financial support from the Pentagon, it would be irrational for them to support the work of young critics, such as Everett and David Bohm. Everett predicted that he would not be given opportunities to find a fruitful career in physics, so he found employment with the Pentagon to work on military strategy (controlled nuclear war), and over the years founded various companies in the fields of military strategy and data-processing (see Byrne 2010; Osnaghi et al. 2009). Bohm, persecuted by the McCarthy witch hunt, was fired by Princeton and went into exile in Brazil, where his unorthodox version of the Louis de Broglie pilot wave theory went into oblivion (Freire 2005). Many years later, Bohm's view was intellectually rescued by John Bell (of non-localization in QM fame) (Bell 2004). To continue the story of both Everett on the *outside* of science and Bohm on the *periphery* of science would be beside the main question under discussion: how should have an idealized Republic of Science treated Everett's formulation of QM?

In order to answer this question, a related question should be first asked in regard to Everett's theory of QM: what is the fundamental problem that the various alternative interpretations of QM address? On Popper's view, the central question of the debate around the interpretation of QM is the question of the *epistemological* and *metaphysical* role of the observer in physical systems. In his words:

> I attempt to exorcise the ghost called 'consciousness' or 'the observer' from quantum mechanics, and to show that quantum mechanics is as 'objective' a theory as, say, classical statistical mechanics ... I shall attempt to substantiate my argument in somewhat greater detail, and to state my own understanding of these issues that have plagued quantum theory over the past fifty years, and my own alternative approach. My thesis in this introduction is that the observer, or better, the experimenter, plays in quantum theory exactly the same role as in classical physics. His task is to test the theory. The opposite view, usually called the Copenhagen interpretation of quantum mechanics, is almost universally accepted. In brief, it says that 'objective reality has evaporated', and that quantum mechanics does not represent particles, but rather our knowledge, our observations, or our consciousness, of particles. (Popper 1982, p. 35)

Is Popper correct in his interpretation of the Copenhagen interpretation which, in his view, seems to be both an "epistemological and metaphysical theory" about the nature of the role of the observer or consciousness in determining the reality of physical systems? According to Popper, the Copenhagen interpretation has two components: first, the observer does not merely test theories but also influences physical systems, and second, the theories of QM are not theories of a supposed independent reality but of our consciousness or knowledge or observations.

Does the Copenhagen interpretation say what Popper says it does? Werner Heisenberg's attempt to explain the concept of nature implied by the physics of QM apparently agrees with Popper's reading:

> While in observing everyday objects, the physical process involved in making the observation plays a subsidiary role only, in the case of the smallest building particles of matter, every process of observation produces a large disturbance. We can no longer speak of the behaviour of the particle independently of the process of observation. As a final consequence, the natural laws formulated mathematically in quantum theory no longer deal with the elementary particles themselves but with our knowledge of them. Nor is it any longer possible to ask whether or not these particles exist in space and time objectively, since the only process we can refer to as taking place are those which represent the interplay of particles with some other physical system, e.g., a measuring instrument. Thus, the objective reality of the elementary particles has been strangely dispersed, not into the fog of some new ill-defined or still unexplained conception of reality, but into transparent clarity of a mathematics that no longer describes the behaviour of the elementary particles but only our knowledge of this behaviour ... The atomic

> physicist has to resign himself to the fact that his science is but a link in the infinite chain of man's argument with nature, and that it cannot simply speak of nature 'in itself'. Science always presupposes the existence of man and, as Bohr has said, we must become conscious of the fact that we are not merely observers but also actors on the stage of life. (1958a, pp. 15–16)

Ironically, Bohm, even as a heretic regarding the interpretation of QM, explains similarly the nature of the issue concerning the physical role of the observer in QM, and thereby the fundamental problem situation of QM:

> We must not say that what is in being is just the universe without us. Rather, it is the universe with us in it. We are part of being and therefore when we observe something we may well change it. If this happens to a significant extent then we must remember that being includes us, our minds, our thinking, our actions and so on. Now in this sense Bohr made a very important contribution when he stressed that the observer plays both an active and a passive role—that we are both actors and observers on the stage of life. But this is not in contradiction with the notion that there is a life and there is a stage upon which we can act and observe. (1957, p. 184)

Perhaps coincidentally, one year after Bohm's remarks were published, Heisenberg used very similar language to explain the new role taken on by the physicist in "modern science": "What we observe is not nature in itself but nature exposed to our method of questioning. Our scientific work in physics consists in asking questions about nature in the language that we possess and trying to get an answer from experiment by the means that are at our disposal. In this way quantum theory reminds us, as Bohr has put it, of the old wisdom that when searching for harmony in life one must never forget that in the drama of existence we are ourselves both players and spectators" (1958b, p. 58).

The problem is how the observer and the experimental situation of the observer in the macro-world influence the events in the micro-world. What is the *logic* of the problem here? Is it that we cannot describe the results of the experiment in the micro-world independently of all observers? I don't think that the "epistemological and metaphysical" formulation of the problem reflects a complete understanding of the *logic* of the problem of the interpretation of QM. The *logic* of the problem involves *the problem of finding a framework for relating two different types of physical systems*: first, the physical system of the observer with the observer's experimental situation (colliders, recording instruments, measuring instruments, and so forth) and, second, the physical system of micro-events in the world of sub-nuclear particles. The theoretical physicist David Mermin translates the interpretation problem of QM in terms of how to relate what the observer sees and talks about in the ordinary world and the correlations the observer describes in relating the sub-system of the apparatus with the sub-system of the micro-events or specimen. The interpretation problem arises from the recognition that QM is about correlations at a different level of physical reality than the level of physical reality which we ordinarily inhabit. In his words:

> We did not evolve having to deal with simple two level systems or even complicated atoms. So the only way we can cope with such systems, which evolution did not outfit us to apprehend directly, is to arrange for them to be subsets of larger systems containing subsystems of the kind we do know something about dealing with. We can then learn about the objectively real correlations that exist between the small and the big subsystems, and try to infer the nature of the systems inaccessible to our intuition from how they correlate with the systems we're equipped to deal with. The larger systems are called "classical", and the process of arranging to correlate them with the small systems is called "the measurement process." (1998, p. 559)

The *logic* of the problem is one of how the observer in the physical system *correlates* the readings on the indicators of the apparatus (the physics of the micro sub-systems that physicists investigate with the apparatus) *to* the macro-world or the classical world (where the physicists with their experimental equipment ordinarily reside).

Let me summarize the fundamental *logic* of the problem of QM:

1. The debate over interpreting QM is about the role of the observer in QM.
2. The question (of the debate) is also in physics and not exclusively about epistemology and metaphysics.
3. The question of interpretation is about the physical role of the observer with respect to specific types of physical events.

The question asks, *what is the physical role of the observer with respect to observing and making experiments with quantum effects?* This question is often called the "measurement problem," though per se the problem transcends the mere measurement of position and momentum in experimental situations. As Wojciech Zurek says: "Quantum measurement problem is a technical euphemism for a much deeper and less well-defined question: How do we, 'the observers', fit within the physical Universe?" (2004, p. 121).

Stage Center: The Fundamental Problem

John von Neumann's mathematical formulation of QM acted as both the model and as the challenge for Everett's formulation of QM. Indeed, von Neumann leaves a challenge to the reader of his classic statement of the *Measurement Problem* that both Eugene Wigner and Everett addressed. Here is von Neumann's challenge where the challenge involves discussing a physical system (I), measuring equipment (II), and an observer (III). The state of the physical system (I) becomes determinate only when the reading of the measuring equipment (II) is recorded—even if only on the retina of the observer (III). Now, suppose there is a "four-term example ... affected by a second observer ... [the analysis] can also be carried out in this fashion. It is left to the reader" (1955, p. 445). Wigner may have been the first reader of von Neumann who took up the challenge of "the [generic] reader." *Wigner's Friend Paradox* was out in print in 1961, after Everett's first version of his dissertation (1956) where we can find *Everett's Observer's Paradox*. The words of Everett found in his notes as a student could have been understood as the first written version of *Wigner's Friend Paradox*:

> We consider an observer O(1) in an isolated room about to make a measurement on system S(1) in the room, whose wave function Ψ (S) is known to him. Subsequent to the measurement he will write the result of his measurement on the blackboard B. Outside the room, and isolated from it is a second observer O(2) who is in possession of the wave function Ψ (1) of the system S′ which consists of the entire room containing S, O(1), and his apparatus. O(2) will not interact with S(1) for one year, at which time he will look at the blackboard to determine the result of O(1)'s measurement. He computes from his wave function what he shall see on the blackboard, in which case O(1) was deluding himself when he ascribed an objective probability to the event, or else he cannot predict what is on the board, in which case his probabilities are subjective, since the writing on the board is a classical thing. (Barrett and Byrne 2012, p. 15)

For the sake of completeness, here is Wigner's own description of his paradox:

> What is the wave function if my friend looked at the place where the flash might show at time *t*? The answer is that the information available about the *object* cannot be described by a wave function. One could attribute a wave function to the joint system: friend plus object, and this joint system would have a wave function also after the interaction, that is, after my friend has looked. I can then enter into the interaction with this joint system by asking my friend whether he saw a flash. If his answer gives me the impression that he did, the joint wave function of friend + object will change into one in which they even have separate wave functions (the total wave function is a product) and the wave function of the object is $\Psi(1)$. If he says no, the wave function of the object is $\Psi(2)$, i.e., the object behaves from then on as if I had observed it and had seen no flash. However, even in this case, in which the observation was carried out only when some information (the *yes* or *no* of my friend) entered *my* consciousness. It follows that the quantum description of objects is influenced by impressions entering my consciousness. (1967, p. 176)

Though there is a variation from the version of Wigner's Friend Paradox that Everett heard from Wigner while lecturing, and the version that Wigner ultimately put into print, the point is the same: activities of observers in the classical or macro-world affect micro-events, and present actions of observers are affected by future activities of observers who are observing the observers.

Everett, in the long unsubmitted version of his Ph.D. thesis of 1956 (replaced in 1957 at Wheeler's direction with a shorter, less contentious version regarding both von Neumann and the Copenhagen schools of thought) cuts the Wigner's Friend *Paradox* at its core and shifts the traditional interpretation problem in a new direction. It does so in the following thought experiment, where the traditional way of posing the interpretation problem of QM is transformed from a problem of pure physics into a problem of cybernetics and mathematical game theory. In Everett's words:

> Isolated somewhere out in space is a room containing an observer, A, who is about to perform a measurement upon a system S … In the meantime, however, there is another observer, B, outside the room, who is in possession of the state function of the entire room, including S, the measuring apparatus, and A, just prior to the measurement. B is only interested in what will be found in the notebook one week in the future … One week passes … At this point, B opens the door to the room and looks at the notebook

> (performs his observation). Having observed the notebook entry, he turns to A and informs him in a patronizing manner that … A owes his present objective existence to B's generous nature which compelled him to intervene on his behalf. However, to B's consternation, A … rudely punctures B's ego by observing that if B's view is correct, then he has no reason to feel complacent, since the whole present situation may have no objective existence, but may depend upon the future actions of yet another observer. (Barrett and Byrne 2012, pp. 74–5)

In his variation on the *Wigner's Friend Paradox*, Everett's point is that in order to avoid the absurdity that the activities of observers depend upon the future activities of other observers, we need to resolve the following problem situation for QM as a theory of physical systems: first, the universe of physical systems must allow for multiple observers (internal to each physical system); second, each observer can observe each other making observations; third, each observer's observations of their own physical system must have an existence or reality independent of the other observer's observations; fourth, the system for describing the observations of the other observer must also apply to the observer who is observing the other observer's observations; and fifth, the system for describing the observations of an observer, if relative to the observer's frame of reference, must be relative in the same manner to all other observers' frames of reference.

How does Everett resolve the problem situation presented by his version of Wigner's Friend Paradox? According to the major gravity theorist, Bryce DeWitt (1970), who was among the first to take up Everett's theory, Everett proposed a multiverse theory where every observer has infinitely multiple clones in distinct universes. But there is disagreement about DeWitt's popular interpretation of Everett. Jeffrey Barrett (1999) has a strictly mathematical-empirical reading of Everett that is neutral toward all physical interpretations of Everett's mathematical formulation of QM. There are other physical extrapolations from Everett's mathematical formulation afloat that I will mention later. However, here I propose a straightforward physical extrapolation: Everett developed a single universe theory that is composed of a single universal wave function where observers and observers observing each other form wave functions as *relative states* to the universal wave function.

Everett used information theory, cybernetics, and game theory (1957a) as models for approaching von Neumann's *Measurement Problem*, Wigner's *Friend Paradox*, and the *role of the observer in physical systems.*

For Everett, the observer is not outside the physical system as assumed in both classical and the Copenhagen interpretation, but is part of the physical system as a type of servo-mechanism in a recursive feedback loop. The recursive feedback loop as part of an overall physical system is only a "slice" or "branch" (using Everett's own terminology) of the universal wave function. Moreover, an observer exists as a physical system relative to the singular universal wave function (1957b). On this interpretation of Everett, where the observer is a recursive feedback loop within quantum systems relative to the universal wave function, there is only one reality—the universal wave function. Observers and their observations are wave functions relative to the singular wave.

Bohm similarly characterizes Everett's interpretation: "what we are dealing with is not a many-universe theory, but a theory of one universe with many viewpoints." He continues: "Everett … seems to regard this notion of relative state as a fundamentally new concept comparable to Einstein's regarding the meaning of a position coordinate as being relative to some material system which serves as a frame of reference … As there are many permissible coordinate frames in relativity so, there are many observer 'frames' in quantum theory" (1993, p. 299). Everett, indeed, could not be clearer that he is proposing a single universe theory: "…'physical reality' is assumed to be the wave function of the whole universe itself…" (Barrett and Byrne 2012, p. 70).

Stage Right and Stage Left: Three Types of Objections

First Type: Dogmatic Acceptance of the Multiverse Theory. David Wallace would reject the view that Everett has a single universe theory. According to Wallace, the multiverse theory of the universe is both true and equivalent to QM: "We have excellent evidence for quantum mechanics, therefore we have good reason to believe quantum mechanics is approximately true; quantum mechanics entails there are many worlds; so there are" (2012, p. 372). Wallace is saying that Everett's formulation equals the multiverse, not theory of, not interpretation of, but the multiverse of QM; and that is the truth about reality (as we know it today). In other words, no other interpretation can be true because Everett's multiverse is the truth.

Second Type: Dismissal or Dogmatic Rejection. The objections of Mario Bunge (in private correspondence with me), Bas C. van Fraassen, and Andrew Whitaker that Everett's interpretation as the multiverse interpretation is in left field, not, at the moment at least, worth taking seriously. According to Bunge, the multiverse theory is science fiction because it is anthropocentric and violates the conservation of energy law. For van Fraassen, the multiverse interpretation is not needed: "Quantum theory is well understood. How can I say this when there are several mutually contradictory, and perhaps equally, plausible, interpretations in the field? Well, I assert it exactly because we do have interpretations that so far stood the test of debate. Perhaps ... Everett's ... which I have not discussed, will eventually prove far superior in ways we don't yet appreciate. But that is not needed, as far as I am concerned, to call quantum mechanics a well understood theory" (1991, p. 336). In other words, the multiverse interpretation is assumed to be the only interpretation of Everett, and the multiverse theory is either silly or not needed. The objection of Whitaker starts with a single universe interpretation of Everett: "For Everett the various branches of the wave function co-exist (in some way)" (2006, p. 292). However, Whitaker thinks that there is no need to discuss the single universe version of Everett because the DeWitt interpretation of the multiverse is the one that has taken off. But the multiverse theory is absurd. Hence, according to Whitaker, Everett can be dismissed from the stage of critical discussion, because the popular interpretation that is front and center stage is absurd.

Third Type: Objections from an Alternative Viewpoint. Bohm takes Everett to be proposing a single universe theory and takes it seriously as an alternative theory to his own version of the pilot wave theory as a form of a hidden variable theory. Though Bohm is careful to note and discuss the varieties of interpretations derived from Everett, including Everett *in the plain*, the DeWitt (and the later improved David Deutsch) many-worlds version, the many-views version, and the many-minds version are attempts to explain the main novel conception of Everett where relative *wave functions or relative states are branches with respect to the universal wave function.* In Wheeler's attempt to explain Everett in the article following Everett's only publication on his theory, Wheeler highlights the novelty of Everett's conception: "The concept of relative state does demand a totally new view of the foundational character of physics" (1957, p. 465). Everett's new conception, according to Bohm, introduces an unnecessary degree of complexity where all the versions of Everett "require new principles and assumptions of a speculative nature going beyond the current quantum theory and beyond our present knowledge of awareness [even as a cybernetic-mechanism] and its possible relationship to Hilbert space" (Bohm and Hiley 1993, p. 315).

Summing up the three types of typical objections to Everett's theory: first, the multiverse interpretation is accepted dogmatically as the truth about physical reality without considering the possibility of a single universe interpretation of Everett nor considering the issues of empirical testability; second, the multiverse interpretation is rejected as needless or absurd where no other interpretation or the issue of empirical testability is considered; and third, the singular universe interpretation of Everett is apparently taken seriously, but the main criticism made by Bohm is that "new principles and assumptions of a speculative nature" (Ibid.) are required by all the varieties of Everett due to Everett's (1957c) *novel theory* of the *relative state*. So, Bohm rejects all varieties of the Everett formulation of QM on non-empirical grounds without consideration of empirical testability.

Back to Stage Center: An Answer to the Objections

Here is a straightforward answer to the three types of objections to Everett: rejecting an alternative approach to QM on an exclusively abstract and non-empirical basis is not putting the theory to a test according to the *methodological and social conventions of the Republic of Science*. Everett's argument for considering his theory as an alternative theory of QM goes as follows: Science produces hypotheses that can be refuted through conceptual testing for logical consistency and empirical testing against experience by deriving predictions. Everett explains:

> One is thus free to build a conceptual model of the universe, which postulates only the existence of a universal wave function which obeys a linear wave equation. One then investigates the internal correlations in this wave function with the aim of deducing the laws of physics ... For example, the classical mechanics of a system of massive particles becomes a law which expresses the correlation between the positions and momenta (approximate) of the particles at one time with those at another time. All statements about subsystems then become *relative* statements. (Barrett and Byrne 2012, p. 158)

Where do observers and observation fit into this model of the universe where the laws of physics are deductions from the universal wave function? Everett answers that observers "can be conceived as automatically functioning machines (servomechanisms) possessing recording devices (memory) and which are capable of responding to their environment. The behavior of these observers shall always be treated within the framework of wave mechanics" (Ibid., p. 77). Moreover, and this is where, according to Everett, we come to the *crucial empirical method for testing* Everett's universal wave theory as well as other general models and theoretical systems for Quantum mechanics:

> A crucial point in deciding on a theory is that one does *not* accept or reject the theory on the basis of whether the basic world picture it presents is compatible with everyday experience. Rather, one accepts or rejects on the basis of whether or not *the experience which is predicted by the theory* is in accord with actual experience ... Thus, in order to decide whether or not a theory contradicts our experience, it is necessary to see what the theory itself predicts our experience will be. (Ibid., pp. 253–254; italics in the original)

Everett is saying that there is only one universe that is completely described by the singular universal wave function. All the laws of physics fall out of the singular universal wave function. So, from the perspective of Jarvie's theory of science as a *Republic* that can, *in theory*, emulate the idealized *Open Society* with *Universal Socratic Dialogue*, we need to reopen the critical discussion of Everett's formulation of QM to test for logical consistency and for whether its predictions are false or not. Everett's formulation of QM has now come out from the cold (now that the Cold War is long gone) and has become a hot topic among philosophers, historians of science, and mathematical-theoretical physicists. Experimental physicists, by and large, are unfortunately still avoiding Everett's formulation of QM (e.g., Saunders et al. 2010). The few experimentally minded physicists who have championed Everett's formulation have adopted updated versions of DeWitt's multiverse formulation, such as David Deutsch (1997, 2016) and Max Tegmark (2004). Both argue for the testability of their multiverse versions of Everett under Popper's theory of the logic of scientific testability. However, to my knowledge, no one is using Everett's own single universe with a singular universal wave function model as a way of developing empirical tests for QM; no one is attempting to derive new predictions so as to discover novel phenomena and find whether those predictions are "isomorphic" to experience and experimentation. In Everett's words: "The theory *is* isomorphic with experience when one takes the trouble to see what the theory itself says our experience will be" (2012, p. 255).

The questions of the standing of Everett's formulation in Big Science and the standing of Big Science in general society are still open: has Big Science become so tyrannical that independent thinkers, such as Everett, are sidelined? Has Big Science, as a social institution or *social technology* that is supposed to search for truth, gone beyond the democratic control of our wider and apparently liberal democratic society? One can be hopeful, along with Jarvie: "If one sets aside reverence for tradition and authority, especially authority that inculcates hierarchy and suppresses nonconformity, then critical and democratic control of science by the wider society seems reasonable enough" (2001, p. 217). He continues, "The present prominence of science, and our dependence upon its technological fallout ... [makes] Science ... too important to escape the twin necessities of structural subordination to the governance of the wider society, and adopting an internal structure of governance that assures accountability within and without" (Ibid., p. 222). Otherwise, the tyranny in science toward its independent thinkers will not cease: "the Everett affair seems to support the idea that as late as in the 1950s the Copenhagen school still exerted a decisive influence, which could go as far as undermining the career of a brilliant physicist in the US" (Osnaghi et al. 2009, p. 121). The story of Everett is not as simple as the collaborators of Bohr acting as Bohr's henchmen in killing off the work of independent scientists as some commentators would like us to believe. Wheeler, Everett's mentor, pushed Everett to make a wave in the Bohrian circle. After Everett failed to gain their approval, Wheeler distanced himself from Everett. Wheeler only showed minimal interest in helping Everett return to physics. Wheeler devised a plan for a working group in the foundations of QM that Everett could join if the plan were implemented. As Eugene Shikhovtsev reports, "Nothing came of this plan, and it may have added a pinch of salt to an old wound" (2003, p. 19). Wheeler made this feeble attempt on Everett's behalf only many years after DeWitt flashed a spotlight on Everett's work by propounding a multiverse version of Everett. But the Everett story is only a ripple in the Republic of Science, as Jarvie might aver, compared to the tsunami wave flooding society by the uncontrolled "technological fallout" of science due to the lack of "governance of the wider society" over the tyranny in science.

References

Agassi, Joseph, and Abraham Meidan. 2016. *Beg to Differ: The Logic of Disputes and Argumentation*. Netherland: Springer.

Barrett, Jeffrey A. 1999. *The Quantum Mechanics of Minds and Worlds*. Oxford: Oxford University Press.

Barrett, Jeffrey A., and Peter Byrne, eds. 2012. *The Everett Interpretation of Quantum Mechanics: Collected Works 1955–1980 with Commentary*. Princeton, NJ: Princeton University Press.

Bell, John S. 2004. *Speakable and Unspeakable in Quantum Mechanics: Collected Papers on Quantum Philosophy*. Cambridge: Cambridge University Press.

Bohm, David. 1957. Remarks Made in the Discussion During the Sixth Session. In *Observation and Interpretation in the Philosophy of Physics: With Special Reference to Quantum Mechanics*, ed. Stephan Körner and M. H. L. Pryce, 171–186. Proceedings of the Ninth Symposium of the Colston Research Society held in the University of Bristol, April 1–4, 1957. New York: Dover Publications.

Bohm, David, and Basil J. Hiley. 1993. *The Undivided Universe: An Ontological Interpretation of Quantum Theory*. New York: Routledge.

Byrne, Peter. 2010. *The Many Worlds of Hugh Everett III: Multiple Universes, Mutual Assured Destruction, and the Meltdown of a Nuclear Family*. Oxford: Oxford University Press.

Deutsch, David. 1997. *The Fabric of Reality*. London: Penguin Press.

———. 2016. The Logic of Experimental Tests, Particularly of Everettian Quantum Theory. *Studies in History and Philosophy of Modern Physics* 55: 24–33.

DeWitt, Bryce S. 1970. Quantum Mechanics and Reality. *Physics Today* 23 (9): 30.

Everett, Hugh, III. 1957a. Recursive Games. In *Contributions to the Theory of Games*, ed. H.W. Kuhn and A.W. Tucker, 47–87. Princeton: Princeton University Press.

———. 1957b [1973]. The Theory of the Universal Wave Function. In *The Many-Worlds Interpretation of Quantum Mechanics*, ed. Bryce S. DeWitt and Neill Graham. Princeton: Princeton University Press.

———. 1957c. 'Relative State' Formulation of Quantum Mechanics. *Reviews of Modern Physics* 29: 454–462.

van Fraassen, Bas C. 1991. *Quantum Mechanics: An Empiricist View*. Oxford: Clarendon Press.

Freire, Olival, Jr. 2005. Science and Exile: David Bohm, the Cold War, and a New Interpretation of Quantum Mechanics. *Studies in History and Philosophy of Modern Physics* 36 (1): 1–34.

Heisenberg, Werner. 1958a. *The Physicist's Conception of Nature, Hutchinson Scientific and Technical*. London: Hutchinson.

———. 1958b. *Physics and Philosophy: The Revolution in Modern Science*. New York: Harper & Row.

Jarvie, Ian C. 2001. *The Republic of Science: The Emergence of Popper's Social View of Science, 1935–1945*. Amsterdam and Atlanta: Rodopi.

Mermin, N. David. 1998. The Ithaca Interpretation of Quantum Mechanics. *Pramana* 51: 549–565.

Neumann, John von. 1955. *Mathematical Foundations of Quantum Mechanics*. Princeton, NJ: Princeton University Press.

Osnaghi, Stefano, Fábio Freitas, and Olival Freire Jr. 2009. The Origin of the Everettian Heresy. *Studies in History and Philosophy of Modern Physics* 40: 97–123.

Popper, Karl R. 1945. *The Open Society and Its Enemies*. London: G. Routledge & Sons, Ltd.

———. 1957. *The Poverty of Historicism*. London: Routledge & Paul.

———. 1959. *The Logic of Scientific Discovery*. London: Hutchinson.

———. 1965 [1963]. *Conjectures and Refutations: The Growth of Scientific Knowledge*. New York: Basic Books.

———. 1982. *Quantum Theory and the Schism in Physics: From the Postscript to the Logic of Scientific Discovery*. Ed. W.W. Bartley, III. London: Hutchinson.

Saunders, Simon, Jonathan Barrett, Adrian Kent, and David Wallace, eds. 2010. *Many Worlds? Everett, Quantum Theory, and Reality*. Oxford and New York: Oxford University Press.

Shikhovtsev, Eugene. 2003. Biographical Sketch of Hugh Everett, III. Accessed December 24, 2017. http://space.mit.edu/home/tegmark/everett/everettbio.pdf.

Tegmark, Max. 2004. Parallel Universes. In *Science and Ultimate Reality: Quantum Theory, Cosmology, and Complexity*, ed. J.D. Barrow, P.C.W. Davies, and C.L. Harper, 459–491. Cambridge: Cambridge University Press.

Wallace, David. 2012. *The Emergent Multiverse: Quantum Theory According to the Everett Interpretation*. Oxford: Oxford University Press.

Wheeler, John A. 1957. Assessment of Everett's 'Relative State' Formulation of Quantum Theory. *Reviews of Modern Physics* 29 (3): 463–465.

Whitaker, Andrew. 2006. *Einstein, Bohr, and the Quantum Dilemma: From Quantum Theory to Quantum Information*. Cambridge: Cambridge University Press.

Wigner, Eugene. 1967. *Symmetries and Reflections: Scientific Essays of Eugene P. Wigner*. Bloomington and London: Indiana University Press.

Zurek, Wojciech H. 2004. Quantum Darwinism and Invariance. In *Science and Ultimate Reality: Quantum Theory, Cosmology, and Complexity*, ed. J.D. Barrow, P.C.W. Davies, and C.L. Harper, 121–137. Cambridge: Cambridge University Press.

CHAPTER 19

Jarvie's *Rationalitätstreit*

Paul A. Roth

In what follows, I examine one ongoing theme in Ian C. Jarvie's attempts to specify sources of rational disagreement, and suggest that identifying those conditions under which a belief should be relinquished proves less clear-cut than he seems willing to acknowledge. In particular, I distinguish between that I will term being Popperian—acknowledging what evidence would count against a belief and when it does—and being Popperish, resorting to attacking those who offer arguments against views that one does not wish to change or relinquish. In this regard, one discerns in Jarvie's writings a tendency to be more Popperish than Popperian, at least with regard to his belief that Popper had things right about conditions of falsifiability.

In a recent article, Jarvie bemoans and puzzles over (and not for the first time) what he perceives as a lack of uptake of Popper's thought: "The scope and depth of Popper's philosophy of the social sciences is not much appreciated beyond the small circle of specialists in his ideas" (Jarvie 2016, p. 284). Jarvie maintains that Popper's "intervention in the centuries-old discussion about the nature of human rationality is … his most fruitful and original contribution to contemporary thought about how to grow our knowledge of society" (Ibid.). However, Jarvie proves oddly insensitive to

P. A. Roth (✉)
Department of Philosophy and Jewish Studies, University of California, Santa Cruz, CA, USA
e-mail: paroth@ucsc.edu

R. Sassower, N. Laor (eds.), *The Impact of Critical Rationalism*,
https://doi.org/10.1007/978-3-319-90826-7_19

well-rehearsed tensions in the Popperian views he cites and seemingly endorses. For example, Jarvie records without elaboration that "Popper, like Hayek, opposes holism in moral political deliberations: terrible things were done to real people in the name of collectivities" (Ibid., p. 289). Yet, just a few lines later, Jarvie states that Popper's theory "of objectivity is social: to the degree that it exists, objectivity resides in the institutional framework of science that it is designed to check those biases of individuals that infect their scientific results" (Ibid.). He also acknowledges that Popper admits "that scientific values such as truth, relevance, simplicity and so forth are deeply anchored in extra-scientific value judgments, including religious ones" (Ibid., p. 311). But how then to prize the unwanted holism of the moral and the social/holistic nature of objectivity? Jarvie offers no hint of explication here.

The question of how to distinguish the moral and the scientific, the social and the objective, now appears quite muddied. Jarvie insists, on the one hand, "Popper developed a new theory of rationality: an idea is rational to the extent that it is open to criticism. This was his major breakthrough" (Ibid., pp. 298–299). This statement may well be correct. But, on the other hand, Jarvie also remarks, "Popper adds little to the core notion of instrumental rationality found in Weber" (Ibid., p. 305). Complications come thick and fast, much faster than clarifications or solutions. In this essay Jarvie finally turns to "redeem my claim … that one of Popper's mot original contributions to philosophy of the social science is his rehabilitation of rationality" (Ibid., p. 307) and offers no more than the following bromide: "His originality lay in exporting rationality from the mind to the attitudinal, from the computational to the institutional" (Ibid., p. 308). But this simply returns one to the problem of separating science and the social, one that Jarvie nowhere suggests how to solve. While the problem is not Jarvie's alone, his remarks here prove symptomatic of a Popperish refusal to revise or develop Popperian views in the face of unpleasant facts or robust criticism.

In order to more fully appreciate the ironic reluctance of a Popperian to revise or surrender a belief when confronted with uncooperative facts, I discuss an unpublished paper (Jarvie 1999) in which Jarvie helpfully distills his understanding of scientific objectivity. There he rightly emphasizes, in accordance with both the letter and the spirit of Popper, how ideally a science should have an anti-authoritarian approach to received knowledge, that is, to encourage and allow criticism, and so to take "observation and experiment" as the arbiter of controversies. Science as an institution, he suggests, has been molded to enshrine just those features that allow for open and public challenges to accepted beliefs. Of course, these features do not themselves emerge from a void. They result from reflection on the fact that the emergence of contemporary science seems like the most effective way to gain stable control of our environment. Focusing on what seems to provide the engine for success in science centers on taking natural science as being the best case for promoting an instrumental conception of rationality. Relatedly, testing against experience provides a public and objective standard for assessing scientific claims. With regard to these specific aspects, Popperian rationality becomes Weberian, just as Jarvie states. Understanding the goal of inquiry instrumentally, in terms of prediction and control, any failure to achieve this should be an accepted test of a claim to correctness. The value thus attached to science, including taking it to be a type of standard of good reasoning, itself proves *contingent* on the assumption that specific scientific practices prove the best path to stable, usable knowledge of the world.

In this regard, two thought experiments prove germane. One I first heard proposed by a former colleague of mine, a person who (in my opinion, anyway) possessed outstanding analytic skills and who was an Orthodox Jew. I forget now how the discussion evolved, but it concerned the rationality of religious belief. I expressed deep skepticism about the very possibility of justifying a belief in a supreme being. (Jarvie, in his writings, has also expressed similar skepticism. Indeed, religious belief often serves for him as a ready example of rationality gone awry.) My colleague challenged me, in a Popperian spirit, how I might react if when staring at the heavens one starry night, the stars were to re-arranged themselves into a message that read, "Hello, Paul, this is God contacting you." My initial response was that there had to be some other explanation for what I observed. But he pressed the point, and rightly so. The discussion forced me to acknowledge that on the one hand, although I am inclined to say, "I find no evidence for the existence of a supreme being," on the other hand, I proved quite unable to imagine an experience that could falsify or lead me to alter my belief that no supreme being exists. For any such supposed candidate that would putatively challenge my skepticism in this regard, I came to appreciate that some alternative, non-theistic explanation would invariably be preferred. So, I had to admit to having a belief that I took in some sense to be evidence-sensitive, but which I also could not acknowledge possibly being overthrown by experience. This thought experiment provides a type of template for assessing in oneself what beliefs are or are not evidence-sensitive.

In a related, but less dramatic vein, Willard Quine notes that empiricism and the other rationality making characteristics associated with natural science themselves emerge as a value from *within* science as now understood. Popperians seemingly agree. But Quine goes on to admit that should one find that, for example, Ouija boards prove a better predictor of experience than does physics, instrumental rationality would dictate going with Ouija boards. Presumably Jarvie with his Weberian conception of rationality should also agree. These two thought experiments suggest that if one defines rationality instrumentality, then the values that Jarvie identifies as hallmarks of rationality themselves prove *contingent* on a certain conception of what makes for success in science, one that experience might possibly overthrow. Put another way, taking falsifiability as a test of belief assumes that falsifiability promotes instrumental success. Commitment to falsifiability, in this respect, should be contingent on justifying its utility towards promoting that end. Problems can arise, as the first case discussed above reveals, when one discovers an inability to allow that experience *could* successfully challenge a belief (or non-belief, as the case may be). My concern, then, is whether or not Jarvie himself proves true to an *instrumental conception of rationality* or whether his writings betray a tendency to fetishize the Popperian letter on rationality and so lose the Popperian spirit that he claims to value. In short, when it comes to what makes for the standard of rationality by way of belief, does he prove more Popperish than Popperian?

His writings display some troubling ambivalence here. The ambivalence can most readily be discerned when Jarvie faces a thinker such as Thomas Kuhn, a person who notoriously muddies the waters regarding just what counts as empirical or as an experimental "refutation" of a theory or a related hypothesis. Kuhn (2012) offers a descriptive account of how science develops, not a normative account. A key problem that famously arises for Popperians from Kuhn's descriptive account involves the fact that from a Kuhnian historical perspective, the usual Popperians' measures of scientific rationality, such as empirical falsifiability, become problematized and so lose their status as *independent* markers of what to rationally reject. Jarvie's reaction to such a Kuhnian history displays unfortunate similarities to my reaction to my colleague's Popperian challenge to my anti-theistic views. That is, Jarvie's response betrays the fact that seemingly nothing could falsify literal Popperian views about what constitutes rational behavior with respect to tests in science.

Jarvie's commitment to taking whatever promotes instrumental success as the mark of the rational turns out to be not Popperian in spirit, but rather Popperish with respect to those who argue against the letter of the Popperian doctrine. In this key respect, Jarvie's views on rationality prove quite *unlike* Quine's, as discussed in the second thought experiment, insofar as Quine acknowledges science as a contingent standard of rational inquiry. Jarvie clearly but doubtless unwittingly reveals this by essentially presenting an *ad hominem* attack against those inclined to promote or to accept Kuhn's views, including Kuhn himself. Jarvie obscures the Popperish and so non-Popperian nature of his response by distinguishing between a philosophical thesis and a sociological one:

> My *philosophical* thesis will be that sociology engages in illicit reduction if it tries to explain socially the truth or untruth of scientific ideas. My second, *material or sociological*, thesis will be that the ideas Kuhn and Popper put forward are essential to explain … their differential "presence" to the sociology of science … Kuhn's ideas legitimate science's current social embodiment; Popper's undermine it … The American academic and scientific Establishment, sensing these unintended consequences, has naturally preferred to embrace Kuhn and hold Popper at a distance. (Jarvie 1988, 301)

As it turns out, Jarvie spends little space or philosophical effort with regard to what he labels his "philosophical thesis." Indeed, he only reiterates the well-rehearsed literal Popperian line on this matter: success in science consists "in refutable assertions about the world and such social formations as would foster their production and refutation" (Ibid., p. 305). But if the sciences themselves determine what counts as successful refutation as opposed to somehow straightforwardly responding to some independent determination of such factors, then what Jarvie terms the "intellectualist" distinction between a sociological and non-sociological view of science blurs. Put another way, if a Kuhnian history of science can be taken to establish that changes in theoretical style have *not* been driven, as Popper imagines they ought to have been, then this imperils attempts to locate why science succeeds in a conception of falsification that stand above or apart from the sociological fray. The institutions of science, rather than securing freedom of experimental results from unwanted influences, work rather (as some read Kuhn) as often as not to insulate received theory against such disconfirming results.

What, then, of what Jarvie terms his "material or sociological" thesis, that is, that ideological concerns tied to an intellectual "Establishment" accounts for Kuhn's ascendance, on the one hand, and Popper's alleged neglect, on the other? In particular, Jarvie proffers the following criticism of Robert Merton's analysis of Kuhn's acclaim:

> He [Merton] fails to consider that the vested interests of the system are such that whatever its creatures produce will be hailed as major new achievements. And here is where the self-exemplifying system character of the sociology of science breaks down. Science offers something that might be naively described as conceptions of nature. These conceptions, moreover, can be tested against evidence other than that considered by their creators. What, however, can play the role of independent check on the sociology of science? (Ibid., p. 309)

Now, as a challenge to claims of sociological hegemony, Jarvie's point cannot be slighted or ignored. (I have myself referred to such sociological accounts as "voodoo epistemology" (Roth 1987).) But the issue at hand does *not* concern how such sociological claims might self-reflexively render themselves problematic (including, as discussed below, Jarvie's own recourse to such sociological arguments). Rather, the hypothesis to be put to the test concerns Jarvie's proposed explanation of the "differential presence" as Jarvie estimates it with respect to reasons why Kuhn's account of scientific change prevails over that of Popper's, at least insofar as the intellectual "Establishment" has it.

Jarvie stresses that he asks questions not in the spirit of philosophical counterfactuals, but rather "traditional sociological questions, namely, why did these ideas find a suitable home among these persons and institutions at this time?" (Jarvie 1988, p. 310). And here Jarvie enters his own explanatory hypothesis: "Merton quite underestimates 'the Popperian presence,' which is required to explain a great deal of what is happening in philosophy and the natural sciences—even though Popper is rarely mentioned" (Ibid.). On Jarvie's reading of Kuhn, science "advances"—solves problems within the context of what Kuhn terms 'normal science—precisely by denying what Popper so vehemently affirms, viz., that science must function as an open as opposed to a closed society (Ibid., pp. 310–1). "Yet his [Kuhn's] argument is that when it comes to science if we want to make progress ... we have to suppress anarchy and impose intellectual discipline" (Ibid.).[1] This would seem to pose an interesting ***empirical*** hypothesis that could then be challenged by developing an account of a history of scientific change in a way that plausibly falsifies Kuhn's narrative. But one finds no such empirical/experiential challenge to what is, *prima facie*, just the sort of bold hypothesis to received historical dogma a Popperian might be thought to welcome. Whatever else Kuhn's book represents in 1962, it did not repeat the orthodoxy regarding how science changes and evolves over history.

An early appreciative characterization of what role history plays in Kuhn's work appears in Gerd Buchdahl. As he observes, for Kuhn "we must regard the scientific enterprise as somehow fragmented into a number of relatively (temporarily and ideologically) isolated periods," from which "it must follow that the history of ideas, including philosophical ideas, will likewise appear as a series of 'Gestalt-views' in terms of which we shall tend to interpret more special enterprises like those of history in general, and history of science in particular" (Buchdahl 1965, p. 55). Buchdahl puzzles as to why one "must regard" the history of science as "fragmented" into "isolated periods." Yet, as Buchdahl also goes on to note: "There is no doubt that within their context, the main contentions of this book imply a refreshingly new approach to both science and its history" (Ibid., p. 56). Buchdahl, a historian and philosopher of science, exhibits a sensitivity to the fact that any distinction between describing and justifying becomes vanishingly small when doing history: "Reading this volume, one gets the suspicion that the author's facts are used to *illustrate* a preconceived notion of historiography, not to *prove* it. Nor is this necessarily a very damning criticism" (Ibid., p. 59).

In striking contrast, another early review of Kuhn's book by a prominent historian of science maintains that Kuhn "is not writing history of science proper. His essay is an argument about the nature of science, drawn in large part from its history" (Gillispie 1962, p. 1251). But this characterization fails to notice exactly what Kuhn takes to be a core issue, whether anything could (without distortion) count as a "history of science proper," or alternatively, what sense could be assigned to the phrase "science proper." Gillispie thus misses Kuhn's moral, or as I shall say, what he seeks to explain.[2] For in narrative explanations no functional distinction between proof and illustration—the normative and the descriptive—can be drawn. "But I think that after reading this book no historian and no philosopher of science will ever be quite the same again ... Of one thing one can be certain; that we have here a new historiographical paradigm which will surely leave its mark on future generations of historians of science"

(Buchdahl 1965, p. 64). Buchdahl thus appreciates in a way a reader such as Gillispie did not that *Structure* instantiates *qua* history of science the very type of history and so philosophy of science for which Kuhn advocates.

Just at the point where the Popperian letter threatens to conflict with the Popperian spirit, a challenge to empiricism in the name of instrumental rationality, Jarvie's argument turns straightforwardly *ad hominem* with regard to Kuhn: "Dare one see elements of autobiography in all of this? Is Kuhn describing how he, as a young man with a mind that tended to wander, was forced to discipline himself (no questioning of fundamentals) to get his degrees in physics?" (Jarvie 1988, p. 311). To this dismissal of Kuhn, Jarvie adds another alleged shortcoming of Merton's account of Kuhn's acceptance: Merton's account "is too thin on Popper to explain Popper's outsider status and his later transformation into the leading philosopher of science he is now" (Ibid., p. 312). No mention here, alas, of problems regarding the analytic/synthetic distinction or verifiability which positivists recognized and which represented significant *internal* challenges to their views. Indeed, Jarvie's sociological explanation of Kuhn's triumph proves as brief as it is polemical: "Kuhn was recruited as part of a pool of talent whose job was to reinforce the collapsing ideology of an insecure establishment. Hence Kuhn's elevation from author of an obscure (and, some thought, iconoclastic) monograph to a major figure in American academic life" (Ibid., p. 317). Jarvie conjoins this "sociological" claim with a quasi-philosophical one: a "darker secret, one still not acknowledged in public, was that the one person primarily responsible for this ruin [the downfall of logical positivism] more than any other was Karl Popper" (Ibid., p. 316). The reason for this being secret, Jarvie suggests, concerns the fact that Kuhn's account fitted the needs of post-WWII and Popper's did not.[3]

But how credible should one find this narrative? Obvious potential falsifiers to Jarvie's historical hypothesis, such as the other challenges facing positivism noted above, receive *no* mention or consideration. Even worse, Jarvie's own historical hypothesis ill fits much other evidence as well; it remains a philosophical mystery cloaked by a methodological irony, accounting for the logic of explanation that underwrites the influence and status of Kuhn's book (hereafter SSR). What makes this a mystery? Consider the following irony: despite SSR's status, there exists no generally accepted specification of those features that an historical explanation ought to possess. In this key respect, 50 years of debate regarding the merits of SSR has proceeded virtually without mention of the philosophical void regarding the topic of historical explanation. Almost all readers of SSR, including it seems Jarvie, have stared this mystery in the face without taking notice of or commenting on it.[4] Indeed, the fact that the entire topic of historical explanation fell off the map of (analytic) philosophy decades ago compounds the mystery of how SSR could have been influential. In this sense, the mystery runs deep, for it has managed to elude detection even by those supposedly highly sensitized and trained to identify, analyze, and evaluate standards of explanation and argument. Just at a moment when philosophy of history arguably should have taken off as a core philosophical discipline, riding a wave of professional concern one might have expected SSR to generate, the topic effectively disappears.

Why pursue this philosophical cold case, the unsolved and allegedly worrisome mystery of a philosophical topic gone missing? Why worry about the silence that surrounds questions regarding the norms of historical explanation? I suggest the following answer: breaking the silence should prove key to exposing still existing and important questions about the relation of history and philosophy, ones that presently go unasked and ignored in polite philosophical company. As noted, Jarvie fails to ask how a narrative explanation could be falsified. For historical narratives (including why Kuhn prevails) do not rely on notions of verification, such as those debated in Popper's *The Poverty of Historicism* (1963). Regarding the historical/philosophical context at the point when SSR first appears requires situating Kuhn's work relative to Carl Hempel and to Quine. Reading Hempel's classic (1950) alongside Quine's classic (1951), a striking feature emerges: Hempel embraces the type of the holism that Quine so (in)famously promotes. Both acknowledge that holism radically broadens what counts as the unit of empirical significance with regard to explanation and testing in science. Ironies abound here. On the one hand, Hempel betrays no anxiety that holism ultimately represents any principled problem for his favored analysis of the logic of science. On the other hand, Quine hypothesizes that holism makes it impossible to philosophically vindicate the verification criterion of meaning at the philosophical heart of positivism.

Kuhn for his part explicitly acknowledges Quine's critique as a key influence, particularly the holism it ushers into philosophical prominence (Kuhn 2012, p. vi; see also Zammito 2004). Kuhn's particular narrative of a history of science powerfully illustrates how this shift in a conceptualization of the unit of empirical significance effectively upends all prevailing accounts of what supposedly explains the rationality of theory change in science. And this turn of events—the emergence of a narrative of the history of science that profoundly alters and constrains any philosophical account of how rational evaluation of scientific reasoning could proceed—would galvanize philosophical concerns about and research into the nature of historical knowledge and explanation. Yet, as Arthur Danto wryly notes, nothing of the sort happens:

> I can think of very little in the philosophy history from the middle-1960s to the present. Somewhere someone sometime in the last decade must have written about explanation, even about historical explanation—but I cannot think of an example offhand ... It is not just that the topic is under extreme neglect. It is, rather, that there is hardly room in the present scene of philosophy for discussion of its issues. So to find someone actively working at them would be almost to encounter a historically displaced person, like someone doing abstract expressionist canvasses as if the whole subsequent history of art had not taken place. (1995, pp. 72–73)

The spell cast by positivism conjured analytic philosophy of history into existence. Those caught in the magic of that moment perceived a need to exorcise history of its possession by narrative form. But when the positivist spell breaks, such concerns vanish.

Danto, himself a key player in philosophy of history, identifies Kuhn as the thinker who forces philosophers of science to rethink the philosophical role that they must accord to history:

> What makes Kuhn's work *historically* important is the fact that a good many thinkers, whose worlds very largely overlapped Hempel's ... were caused by Kuhn's work to turn into thinkers whose world overlapped Kuhn's world instead ... I can remember one of them saying with a cry of anguish that he wished Kuhn had never written that damned book ... [A 16th century scholastic] said, in much the same spirit, that 'The wretched Luther had emptied the lecture halls.' For a long period, there were questions with which scholastic thinkers dealt and with which everyone who shared their world regarded as of the greatest moment. And then, all at once, almost overnight, nobody cared any longer ... [Hempel's theory] just stopped being relevant, the way the whole philosophy of history it defined stopped being. It was replaced with a different set of questions, a world in effect, into which it no longer fit. (Ibid., pp. 84–85)

But Danto, his sophistication with regard to this topic notwithstanding, nonetheless never pauses to ask why philosophy of history fails to rise from the ashes of positivism: "Kuhn advanced a view of history so powerful that history rather than being an applied science, as Hempel holds history to be, history came to be the matrix for viewing *all* the sciences. It all at once became the philosophical fashion to view science historically rather than logically, as an evolving system rather than a timeless calculus, as something whose shifts over time are philosophically more central to its essence than the timeless edifice of theories" (Ibid., p. 72).

Hempel held that histories only offer "explanation sketches," explanatorily adumbrated accounts that need to be filled out by covering laws in order to achieve full legitimacy as scientific explanations. But the *de facto* impact of Kuhn's historiographic practice proves false to Hempel's attempt to legislate the form of historical explanation. In short, Danto gives voice to the fact that Kuhn's work made passé all that supposedly defined what a science of history had to be. Yet having commented on the surprising result of Kuhn's work—the complete and sudden overthrow of a powerful theory of scientific explanation by means of a work of history that supposedly does not have even *prima facie* status as a scientific explanation—and stared it in the face, Danto (like so many other philosophers) then simply turns away.

The old questions remain unanswered; the demise of positivist hegemony in philosophy of science only removes the pressure to answer them. In short, once positivist-inspired methodological debates cease to have any real point, interest in philosophy of history within analytic philosophy largely disappears. Ironically, subsequent to the publication of the first edition of SSR, as the history of science becomes central in the philosophy of science, philosophical discussion of historical explanation effectively ceases. And even as SSR has been widely noted and much discussed (Zammito 2004), with a distinctive style of sociology of science and inspires the creation of science studies, questions of what makes for a proper historical explanation remain ignored. *SSR effectively reverses the received order of epistemic authority.* Prior to Kuhn, in order for a work of history to count as providing a legitimate explanation, it needed to conform to a certain standard determined by an ahistorical account of science. Post-Kuhn, philosophers fashion histories to account for which explanatory forms come to prevail and why. Jarvie clearly does this as well. But in doing so, he too passes over in silence how one "tests" competing narratives against facts, especially in cases where narratives deviate not with respect to specific matters of fact, but with regard to more nebulous claims regarding intellectual or sociological influence.

Moreover, Kuhn never receives attention as a historiographer or a philosopher of history. Indeed, his remarks on historical methodology prove sporadic and mostly unilluminating. If analytic philosophy of history begins with Hempel's throwing the gauntlet down and daring historians to meet the challenge posed by standards of scientific explanation, it ironically ceases just at the point of a miraculous reversal of fortune. Despite decades of exile from the realm of scientific explanation, a work in the history of science overthrows extant accounts of the rationality of theory change in science. But how could this have happened, given the absence of any accepted basis for judging a history as explanatory? Kuhn certainly thought of himself as a historian, and SSR comes chockablock with historical examples. Yet its well-known opening sentence calls for particular scrutiny in this regard: "History, if viewed as a repository for more than anecdote or chronology, could produce a decisive transformation in the image of science by which we are now possessed" (Kuhn 2012, p. 1). This sentence poses an under-appreciated puzzle. Indeed, a sort of willed blindness persists with regard to construing Kuhn's naturalism specifically in terms of a historiographic method. Alexander Bird, for one, does not even consider historiography in this context: "For a philosophical audience in particular, its approach, which we would now call 'naturalistic' in its use of empirical science, was unusual. Kuhn's arguments, especially as regards philosophical consequences, are often more implied than stated explicitly" (2012, p. 865). Even John Zammito's otherwise sophisticated account negatively assesses Kuhn's work as a history: "This opening sentence offered a clarity of intent the whole work failed to sustain" (Zammito 2004, p. 52).

But clearly Kuhn does not intend the historical cases to function only as counter-examples to then extant philosophical accounts of theory change in science. To assume he does makes it incomprehensible as to why Kuhn might have any reason to expect a "transformation in the image of science." For read only as a motley collection of counter-examples to verificationist or falsificationist models, no *alternative* notion of science follows. Yet from his very first sentence, Kuhn cautions against reading SSR as an exercise in historical bricolage. At the very least, any explanatory logic imputed to SSR must help make plausible how it proposes to affect such a "decisive transformation." Inasmuch as Jarvie would have us reject one narrative and replace it with another, he owes some account of what falsifies a narrative. But he fails to recognize what sort of challenge to received views Kuhn actually poses by providing a particular type of history.

Moreover, Jarvie's suggestion that a chief reason accounting for Kuhn's inclusion and Popper's exclusion must lie in the fact that Kuhn does and Popper does not give aid and comfort of some shadowy intellectual Establishment fails on several points of historical plausibility. First, it ignores completely Quine's fundamental challenge to positivist orthodoxy by way of his criticism of the analytic/synthetic distinction and its implied holism. Second, it overlooks as well how Kuhn's work "taps into" Quine's criticism and exploits it regarding an alternative history of scientific change, one which already had orthodox positivists such as Hempel worried. Kuhn's use of historical method, in turn, further challenges any claims to scientific orthodoxy. Finally, no evidence that Jarvie offers indicates that Kuhn, quite unlike Popper, tried to provide a normative account of science, and so a justification of science as he found it. Jarvie's relentless critiques of other views over the years display a perfectly praiseworthy commitment against countenancing any such blurring of the lines. But one would then expect something more than a Popperish response to Kuhn's bold challenge.

I once heard Richard Rorty remark that it sometimes helps to play the village positivist. But this I understood Rorty to mean that all the well-known and much discussed problems with the verifiability criterion of meaning notwithstanding, it remains a philosophically worthy move to ask at first pass what the truth conditions of a particular claim might be. In a similar vein, much merit attaches to being the village Popperian, and asking what evidence might count against a belief. Playing the village positivist or Popperian could well reveal that if no such evidence could play the requested role, one confronts mere dogma and not a claim that has some actual empirical or scientific license. But, to paraphrase John Austin, being a village positivist or Popperian might be the first word but cannot, philosophically speaking, be the last word. Whether or not Popperians can advance their signature claim about falsifiability as definitive of rationality beyond Popperish responses remains undetermined.

Notes

1. The ellipsis in this quote marks my omission of the words "note his contradiction" that Jarvie has a parenthetical aside in the original. The presumed contradiction, alluded to but not explicitly developed, concerns Kuhn's own presumed commitment to democratic values—including, presumably, valuing the sort of critical exchange that Popper appropriates from Mill—and his

characterization of what makes for scientific success. Again, Jarvie reads into Kuhn's descriptive account a normative recommendation. But no one ever accused Popperians of being charitable interpreters.

2. The choice of Gillispie as a reviewer proves interesting inasmuch as Kuhn cites him by name as authoring a history of science shaped to fit the "Procrustean bed" of progressive development from which Kuhn looks to free this subject (Kuhn 2012, p. 108, fn. 11).
3. Given Jarvie's line of argument in this chapter, it comes as no surprise that he pens an extremely positive, almost gleeful, review of Steve Fuller's attempted deconstruction of Kuhn almost a decade and a half later. Fuller largely retails in greater detail the charges that Jarvie rehearses in his 1988 paper. See Jarvie (2003, pp. 275–83). For a counterpoint of how Fuller tells this tale, see Roth (2003).
4. This paragraph and several that follow have been adapted from Roth (2013).

References

Bird, Alexander. 2012. The Structure of Scientific Revolutions and Its Significance: An Essay Review of the Fiftieth Anniversary Edition. *British Journal for the Philosophy of Science* 63: 859–883.

Buchdahl, Gerd. 1965. A Revolution in Historiography of Science. *History of Science* 4: 55–69.

Danto, Arthur. 1995. The Decline and Fall of the Analytical Philosophy of History. In *A New Philosophy of History*, ed. F. Ankersmit and H. Kellner, 70–85. Chicago: University of Chicago Press.

Gillispie, Charles C. 1962. The Nature of Science. *Science* 138: 1251–1253.

Hempel, Carl G. 1950. Problems and changes in the empiricist criterion of meaning. *Revue Internationale de Philosophie* 4: 41–63.

Jarvie, Ian C. 1988. Explanation, Reduction and the Sociological Turn in the Philosophy of Science: Kuhn as Ideologue for Merton's Theory of Science. In *Centripetal Forces in the Sciences*, ed. Gerald Radnitzky, 299–320. New York: Paragon Press.

———. 1999. Rationality and Situational Logic in Popper's Scientific Work. Unpublished. http://www.yorku.ca/jarvie/online_publications.htm.

———. 2003. Fuller on Science. *Philosophy of the Social Sciences* 33: 261–285.

———. 2016. Popper's Philosophy and the Methodology of Social Science. In *The Cambridge Companion to Popper*, ed. Jeremy Shearmur and G. Stokes, 284–317. Cambridge: Cambridge University Press.

Kuhn, Thomas S. 2012 [1962]. *The Structure of Scientific Revolution*. Chicago: University of Chicago Press.

Popper, Karl R. 1963. *The Poverty of Historicism*. London: Routledge.

Quine, Willard V.O. 1951. Two Dogmas of Empiricism. *Philosophical Review* 60: 20–43.

Roth, Paul. 1987. *Meaning and Method in the Social Sciences*. Ithaca, NY: Cornell University Press.

———. 2003. Fuller's '18th Brumaire of Thomas K'. *Social Epistemology* 17: 281–289.

———. 2013. The Silence of the Norms: The Missing Historiography of *The Structure of Scientific Revolutions*. *Studies in History and Philosophy of Science* 44: 545–552.

Zammito, John H. 2004. *A Nice Derangement of Epistemes*. Chicago: University of Chicago Press.

CHAPTER 20

The Political Philosophy of Science in Historical Perspective: The Road Through Popper and Polanyi to the Present

Stephen Turner

The Enlightenment Reconsidered

One Enlightenment issue is particularly central. Science in the narrow sense is not sufficient for human progress. Carrying out the task of making science benefit society necessarily required a significant degree of knowledge of the social world. Yet this knowledge has always proven to be problematic, difficult to regard as scientific, and in conflict with politics. The difficulties were immediate: in the aftermath of the restoration of French politics as part of the return to normalcy after the revolution in 1803, the section on social and political science of the French academy was suppressed, marking out a line between acceptable science and dangerous science and depriving science of the means of applying itself to the good of society. Social and political speculation then fell to thinkers outside the academy, notably Henri de Saint-Simon, who revived Condorcet's ideas about the relation of science and technology to progress, but reconsidered many Enlightenment ideas, notably the program of secularization, in the light of the Revolution and its failures.

S. Turner (✉)
Department of Philosophy, University of South Florida, Tampa, FL, USA
e-mail: Turner@usf.edu

R. Sassower, N. Laor (eds.), *The Impact of Critical Rationalism*,
https://doi.org/10.1007/978-3-319-90826-7_20

This story is well known and need not be repeated here. Saint-Simon's young secretary, Auguste Comte, revised and extended Saint-Simon's sketchy but illuminating ideas into Positivism, a complete intellectual system and program which provided both a philosophy of science and a model for the relations of science and society as well as a repudiation of liberalism, which Comte and most of the advanced Continental thinkers of the time regarded as a transitory historical phenomenon doomed by its overwhelming defects. For Comte and Saint-Simon, these defects were at the heart of the problem of the relation between science and society and science and politics. Comte, in the most explicit of terms, stated his disgust for the idea that everyone should be permitted to have their opinion heard (1864, IV, pp. 50ff).

The reasoning here is simple, and poses the most fundamental issue: one model of science stresses consensus and authority. This was Saint-Simon's: within science, scientific merit was transparent, and poor scientists would naturally recognize and defer to greatness in others, creating within science a natural hierarchy which was a desirable model for the natural hierarchy of the new scientific and industrial order he envisioned. Authority and consensus within science have implications for what is outside of science. If science is correct, and science is a sufficiently extensive activity to comprise knowledge of such things as the social world, why shouldn't scientists rule, and rule to the exclusion of the ignorant? Is not the rule, *de facto* if not *de jure*, of scientists the condition of progress? And if the understanding and recognition of the authority of science are the central conditions of progress, shouldn't science be imposed on the ignorant as the dogmas of Catholicism had been so effectively imposed in the past, using the same techniques? Comte makes this argument explicit:

> there is no liberty of conscience in astronomy, in physics, or even in physiology, that is to say everyone would find it absurd not to have confidence in the principles established by the men of these sciences. If this is not so in politics, this is due only to the fact that since the old principles have yet to be abandoned, and no new ones have yet been devised to replace them, there are really no fixed principles at all in the meantime. (Ibid., p. 49n)

The main item on the agenda of progress for Comte was to overcome the anarchy of opinions produced by the situation in which the ignorant and the expert were equally empowered to give their opinions.

Of course, freedom of conscience and open discussion is the bedrock of liberalism, so the anti-liberal implications of science as a model for politics are evident. But at the same time, there was, only partly developed, what might be called a liberal conception of science, which is to say one which de-emphasized, even obscured and denied, the issue of the authority of science in favor of a model of free discussion that applied equally to science and politics. James Mill is perhaps the person whose boundless faith in discussion best exemplifies this intuition, but John Stuart Mill's *On Liberty* (1978) is perhaps its greatest expression. But Mill does not mention science in this essay and did not engage directly the puzzle of the existence of scientific "authority" in a non-authoritarian social order. There was, however, a "liberal" critique of the corrupting effects of science's dependence on state patronage in Thomas Henry Buckle (1924), similar concerns about state influence in Condorcet (1976), and protections of the freedom of scientific discourse were written into the rules of the Royal Society. These were primitive elements of an alternative image of science.

Pearson and Mach

The underground influence of Comte's core ideas was substantial, but the most consequential influence was mediated through two thinkers: Ernst Mach, who developed and popularized a philosophy of science which was congenial to certain subsequent developments, notably Logical Positivism, and served as a carrier for some key ideas of Comte's, and Karl Pearson, who had imbibed Comte at Cambridge from a librarian. As social and political thinkers as well as philosophers of science, they had much in common. Both held that the age of coercion or brute force had passed and sought some sort of surrogate for religion. Each was a socialist with an anarchist streak, though they expressed the latter differently. Mach was hostile to the state, at least as presently constituted. Pearson thought that present institutions needed to be "dynamited," but at the same time imagined that the revolution could be made from above by persuasion, and that the aristocracy could be persuaded to part with its wealth. Both saw science as having an integral role in this transformation, and each of them thought that spiritual transformation in the direction of a scientific cast of mind in the public was a necessary part of this. Both thought that Buddhism, as a non-deistic religion, offered a model for appropriate religiosity. And both had a conception of science which minimized and was suspicious of the theoretical and of realism. Each resisted the theoretical innovations of his time: Mach the atom, Pearson the gene. Each was attracted to the idea of efficiency as an organizing concept in science with implications for politics. Both had what might be called an extensive view of science, considering science to be the source of social guidance, to be applicable far beyond the limits of present science, and to be essential to the *Weltanschauung* of the future. They differed on the details of this extensiveness, and Pearson had a much more fully developed view of its political significance.

Pearson's thought and decisive significance come into focus when he is understood both within the context of his own time and in his position as a transitional figure between two bodies of thought widely separated in time, namely, that of Comte and the British Communist theorists of science of the 1930s, such as L. T. Hogben, who recalled that his generation had "been suckled on the *Grammar of Science*" (1957, p. 326, quoted in Porter 2004, p. 7). Pearson is free of Comte's overt hostility to the public expression of opinion, but he is not free of the model of religion. He comments at one point that there is no Pope in science (Pearson 1905, p. 62) and observes that doubt is integral to science and part of science's mystery (Pearson 1937, pp. 50–51). This means for him that a scientific hierarchy persecuting scientific heresy would be fatal to progress. But to say that science has no high priest is of course not to say that it has no religious or quasi-religious significance; indeed, it is to hint at the opposite. Pearson was also greatly concerned with the Comtean question of what should replace traditional religion in an age of science. He was fascinated, as Saint-Simon had been, with the idea of medieval society and the idea of a unified culture, which of course required a new religion. Like Comte's and Mill's religion of humanity, he predicted the deification of man (Pearson 1888, p. 31). He was against "rights," against individualism, for solidarity, and for the state, in principle. Toward present state practice he was hostile: his mode of expressing hostility to liberalism was to denounce the elected officials and bureaucrats for their systematic incompetence and to work for reforms in science education. His underlying statism overcame his expressions of hostility to the state, however, and despite his hostility to brute force, he made an exception for offenders against the state whom he said should be strung up.

Pearson resolved these apparently conflicting impulses under the label "free thought." He hoped for the establishment of "poets, philosophers, and scientists ... as 'high priests'" (Pearson 1888, p. 20) and for the elevation of "reason, doubt, and the 'enthusiasm of the study' above the 'froth' and 'passion' of the 'marketplace'" (Ibid., pp. 130–31, 133). How was this to be accomplished? Through "the guidance of the masses by their educated sympathizers," by brains and numbers, as a contemporary follower of Comte put it (Porter 2004, p. 108). Science, in short, enabled liberal politics, with its froth and passion, to be replaced by the leadership of the scientifically elevated. This presupposed a particular conception of the nature of politics, owed to Saint-Simon, in which the real questions of politics were administrative ones, which, for Pearson, meant they could be handled unemotionally and without bias. One of the functions of statistics was to override subjectivity and produce consensus. The scope of applicability of the scientific method, as he construed it, was not only extensive but almost universal, and especially applicable to political questions. The aim of scientific education was to make citizens who could think in this way. And thinking in this way would make politics without coercion, based on consensus, possible (Pearson 1937, pp. 11–14).

This yields a model of science state relations of considerable complexity, with some intriguing gaps. Science as an activity was itself justified by the fact that "its existence tends to promote the welfare of human society" (Ibid.), meaning, that it produces social stability and social efficiency, the analogue of Comte's order and progress. Even pure science could be justified in this way, because the record shows that it is difficult to predict whether effort in pure science will produce, in the long run, the practical results that would warrant state expenditure on science. But the major role of science is to produce unforced consensus, at least *within* science. The consensus produced by science, of course, must be accepted by citizens, and this is where education and popularization come in, and Pearson gave these topics much thought. He argued that there was an opportunity to upgrade craft training by including the underlying scientific principles of the craft as part of the educational experience. More generally, he was concerned with the right way to inculcate the scientific, unbiased, cast of mind. Mere perusal of works on science, he thought, did not lead to this result: the close study of some small area scientifically, however, did (Ibid., pp. 15–16). He believed that one could expect a high degree of carry-over between this experience of study and the role of the citizen. Pearson was no egalitarian with respect to the hierarchy of scientific talent, however, and the role of the semi-educated citizen was still primarily one of respect for the priests of science, whose consensual judgments on the many areas in which they had no direct knowledge citizens would be bound to accept.

Moreover, there was a problem of epistemic deviance, which demanded strong measures: "The abnormal perceptive faculty (i.e., the kind that failed to arrive at the consensual conclusion assumed to be more or less automatically produced by persons with normally evolved perceptual powers), whether that of the madman or the mystic, must ever be a danger to human society, for it undermines the efficiency of the reason as a guide to conduct" (Ibid., p. 120). Lack of conformity to the canons of legitimate inference, Pearson says, is "anti-social" whether it involves believing "in a sphere in which we cannot reason" or believing on insufficient evidence that a cab will be available at eight o'clock when others depend on us catching a train at half past. These are treated as "ethical" matters, but the ground of ethics is in the end social utility (Ibid., pp. 54–55). Yet the hand of enforcement is never made visible by Pearson, and the stress is on "educational" means of overcoming "superstition" and other deviations of belief.

Pearson's epistemology is also relevant here. For him, the facts of science are perceptual successions, correlations, not the objects of theory, such as genes and atoms. So the idea of arriving at an unforced consensus on these is plausible. What is less plausible is the idea that political questions can be resolved into issues of perceptual succession. Pearson's examples of how political questions can be transformed into questions about correlations included Poor Law reform, about which "the blind social instinct and the individual bias at present form extremely strong factors of our judgment" (Ibid., p. 29). He did not regard these as "valuative" or consider the problem of value pluralism. He thought such questions could be resolved by considerations of national efficiency. The main finding of science that concerned him most was the basic law of heredity that "like produces like." He applied this to the problem of government and national strength by proposing eugenic control over population quality, emphasizing the elimination of the unfit, the racially inferior, and always stressing the necessity for the reproduction of the best minds. This was consistent with socialism, for him, because he believed that socialism required superior persons.

Left Wing Politics and Liberalism

We may call the conception of science of Condorcet, Saint-Simon, Comte, and Pearson an "extensive" conception of science: it is a conception which assumes that something in science, its method, in the case of Pearson; its methodological principle of seeking "positive" laws, in the case of Comte;

or its possession of an overarching principle, in the case of Saint-Simon, could be extended into the realms of the social and political, thus abolishing the necessity for politics, or in Saint-Simon's famous (proto-Marxist) interpretation, replace politics with "the administration of things." Science could be conceived more extensively in a variety of ways: as incorporating technology and engineering, as including the social and mental sciences, as including the policy sciences, even as a source of ethics, a popular theme in the post-Darwinist period. All of these figure in the late Victorian discussion of science. But one form of extension becomes, for the next few decades, especially controversial. The way in which Pearson and later Thorstein Veblen talk about science and engineering, as a cast of mind that carries over from one activity or topic to others, is paralleled, though with significant differences in meaning, by the German notion of *Weltanschauung*. Mach and his successors, including the Logical Positivists and especially Otto Neurath, were interpreted, and sometimes interpreted themselves, as providing a scientific world picture—they did not use the term *Weltanschauung*, to better distinguish their own ideology as authoritative. The quest for a scientific *Weltanschauung* played a role in German thought analogous to the role that the problem of the replacement of traditional religion had played in British and French thought. The problem of whether science could provide a *Weltanschauung* produced a novel issue about the cultural status of science that was highly consequential for what followed in the German-speaking world, and ultimately in the Anglo-American world.

In 1931 an event took place that transformed the discussion of the political character of science. During a congress of history of science in London, a fully developed Marxian account of science was presented. It had been developed in the Soviet Union quite independently of Western Marxism, and sponsored at the highest level of the Soviet ideological apparatus by Nikolai Bukharin, who participated in the volume that presented it. The account was couched in historical rather than philosophical terms, though the two were scarcely separate: Bukharin's own main theoretical work was entitled *Historical Materialism*, which opened with these sentences: "Bourgeois scholars speak of any branch of learning with mysterious awe, as if it were a thing produced in heaven, not on earth. But as a matter of fact, any science, whatever it be, grows out of the demands of society or its classes" (Bukharin 1969, p. 9). The volume of articles applying these ideas had a profound galvanizing effect, especially in Britain. The major point of this text was to show in detailed case studies that science was also the product of the demands of the time for technological results, that the demands were specific to particular social formations and historical situations, and that "theory" was ultimately driven by technological practice. This implied that the idea of an autonomous realm of pure science was a sham and an ideological construction.

One of the central ideas of Marx was that when the conflict between the potentials of the forces of production and the restrictive and limiting class structure and system of economic relations was at its greatest was a revolutionary moment. One of the central ideas of both the fascists and the Soviets was that the rational planning of the economy and cultural spheres was possible and necessary. These ideas had a strong grip on the publics and policy makers of the 1930s, faced with the Depression, in the liberal states as well. In the case of science, a large literature developed on "the frustration of science," the idea that capitalists, incompetent bureaucrats, and liberal politicians stood in the way of the kinds of scientific developments that could overcome the all too apparent failures of capitalism. And there were also two interesting public declarations, one in the late 1920s, another in the early 1930s, by prominent Anglican churchmen, one suggesting a moratorium on science for a decade to allow for a reconsideration of its social consequences, another calling on scientists to consider the social responsibility of scientists.

Out of this situation came a complex, multifaceted Left movement, partly Communist, partly non-Communist, which sought to resolve society's problems through science. John Bernal was the principal figure in this movement and publically embraced Communism. Recall that Condorcet, Comte, and Pearson faced the problem that the condition for their account of science to contribute to progress was that citizens become scientifically educated, if only in a limited way, which required scientists to be, in effect, ideologists, whose ideology was authoritative for the rest of society. This was tantamount to rule by scientists, which even the most enthusiastic Left scientists, such as Bernal, regarded as impractical (though he still regarded it as preferable in principle, raising the question of whether Bernalism was Pearsonism in the vestments of Communism). Communism was not rule by scientists, though it involved an authoritative ideology which regarded itself as scientific. So, it was the closest practical approximation of the ideal of rule by scientists available at the time, and these scientists were deeply interested in the way in which the Soviet Union worked and used science. The leaders of the Marxization of science accommodated the realities of the Soviet Union by arguing first that this was the one country in which science had obtained its proper function, as Bernal put it, and secondly, by taking the view that the Soviet system was benign (all that the phrase "the dictatorship of the proletariat" meant, according to Julian Huxley (1932, p. 3), was that things were administered for the benefit of all). They also argued that neutrality was impossible for the scientist, especially in face of the anti-scientific drive of Fascism, that money for science would flow freely in a rationally organized regime with economic planning rather than markets, and that history was at present in a transitional phase toward a state in which science, understood extensively as implying "a unified, coordinated, and above all conscious control of the whole of social life" (Ibid.), held out the prospect of abolishing the dependence of man on the material world, while taking its rightful role of becoming the conscious guiding force of material civilization, permeating all other spheres of culture.

Science and Communism, in short, were made for each other: the full realization of each required the full realization of the other. There was no conflict between science and Communism, because, as Bernal said, science already is Communism, performing the task of human society, and is doing so in the Communist way, in which "men collaborate not because they are forced by superior authority or because they blindly follow some chosen leader, but because they realize that only in this willing collaboration can each man find his goals. Not orders, but advice, determines action" (1939, pp. 415–16). In practice, as Bernal envisioned it, scientists would be organized into trade unions which would cooperate with other trade unions in producing the five-year plans that they would then carry out.

The Critique of Extensiveness

Bernal and his comrades understood that the issue which made their position unpersuasive to scientists was the notion of planning itself, and the question of what the planning of science would entail. The issues over planning were overshadowed, however, by a discussion of Nazi science in scientific circles that dominated the late 1930s. Nazi science was not only planned, it was extensive in the problematic sense that was pointed to by Lenin's notion that no cultural organization in the Soviet regimes should have autonomy from the party. Under the Nazis, science was expected to conform to Nazi ideology, scientists who were Jews were expelled, and a loud campaign against the Jewish influence in science was mounted. A translation of a paper published in a Nazi journal was published in *Nature* (Stark 1938), and it prompted a huge response in Anglo-American science, cast in terms of "freedom" and the link between scientific freedom and democracy. This provided the initial spur to a discussion of the "autonomy" of science. Bernal responded to this literature by defining the issues in similar terms, as a conflict between freedom and efficiency, which he thought could be resolved within the framework of planning, though this required, as Bernal acknowledged, a redefinition of the notion of "freedom," and a Hegelian notion of the positive freedom of scientists under planning.

The issues of planning, freedom of inquiry, and the autonomy of science produced the complex response that Jarvie elucidates (2001). Michael Polanyi provided an argument for the autonomy of science which was shaped as the claim that science was not in need of political governance in the form of planning, because it was already "governed" sufficiently by its own traditions. Science was a community as distinct from the sort of bureaucratic order that could be made subject to planning. Planning would destroy the feature of community life that made possible the growth of ideas, which was, for Polanyi, the freedom of the scientist to choose which ideas to pursue.

Polanyi's version of this argument addresses the problem of science and democracy in a novel way. If science thus understood is subject to "democratic control," it would not flourish. But science is not an anomaly for democracy. It is similar in character to other communities which are granted autonomy on the basis of their strongly traditional, self-governing character, such as the church and the legal profession. And democracy itself, for Polanyi, is strongly traditional. So the relation between science and democracy should be one of mutual recognition and respect, from one community to another, as well as democracy's interest in the fruits of science, which can best be gained by granting the scientific community autonomy (Polanyi 1951). But autonomy was justified by the specific character of the scientific community *qua* community. James Conant puts this succinctly:

> Would it be too much to say that in the natural sciences today the given social environment has made it very easy for even the emotionally unstable person to be exact and impartial in his laboratory? The traditions he inherits, his instruments, the high degree of specialization, the crowd of witnesses that surrounds him, so to speak (if he publishes his results)—these all exert

> pressures that make impartiality on matters of *his* science almost automatic. (1947, p. 7; italics in the original)

These mechanisms, however, exist only for science proper—not for its extensions into politics, where the scientist has no special claim to objectivity or authority.

Both Conant and Polanyi had a liberal approach to science in the following sense: they thought it was best to govern science indirectly, by facilitating scientists' competition. But Conant, acknowledging the realities of "big science," thought it was necessary to have a set of major elite universities with massive resources in order to make this competition meaningful in the present. Conant's view allowed for intervention when competition failed, and Conant was anxious to provide the conditions for competition. Popper, by stressing the central role of falsification and depicting science as a process of conjecture and refutation, gave a vivid picture of the nature of scientific competition and its underlying character.

Popper could have, but did not, develop an analogy between liberal discourse and scientific discourse, and if he had done so the relations between his account and the Liberal approach might have been more evident. Both science and liberal democratic discussion are limited forms of discourse, governed by a shared sense of boundaries. Popper's way of bounding science, the use of falsification as a demarcation criterion, may have led him to think that there was no need to locate a supporting ethos or tradition. But the difference between the verificationist theory of meaning and falsification serves to place Popper on their side in the question of scientism as a *Weltanschauung*. Verification faces out, so to speak, to those forms of purported knowledge that science might hope to supplant or discredit. It is directed at the larger community. Falsification looks in, to the process of scientific discussion that it regulates, and in regulating makes into a variant of liberal discussion.

The Ghost of Bernal

The Left in Britain had been motivated by the idea of the "frustration of science," the idea that capitalism underinvested in science to the detriment of humanity. This vanished in the Cold War era and the later investment in biotechnology by private capital. Yet the emergence of modern "technoscience" as a phenomenon and the political issues associated with its control make Bernal's questions of the 1930s relevant again, though in an odd way. The Left, in the 1960s, became "participatory" and valued the idea of the superior rationality of science and revolutionary ideology less and popular protest movements more. This meant that the Left welcomed popular protest against science and technology, against expertise, and against the global system of capital. Technoscience was now seen as a part of this system. When the global warming movement developed, however, there was a turn back to science, and to the idea of scientific consensus as an authority superior to politics and liberal democracy, and to the idea that liberal democracy was a fatally flawed system that was incapable of dealing with modern technical reality. This had been Bernal's thesis as well.

Reading Bernal today, one senses the momentous character of this change in the status of science on the Left. One indicator of the change is the fact that Bernal was, unbelievable as this might seem, an advocate of the warming of the Northern Hemisphere by technological means, praising the efforts of the Soviet Union to make its northern regions habitable through climate changes that would have affected the whole hemisphere as a heroic application of science for human good (1939, pp. 379–80). It is perhaps only a grim irony that these efforts were part of the Gulag, and that the camps that the Soviets established, and the rail lines that they built to serve them, were swallowed up by the cold, and succeeded only in serving as a means of killing those who were condemned to work on them (Alexopoulos 2017, p. 153). The reasons for the change are clear enough. When capitalists embraced science for profit and cloaked it in slogans about human betterment, such as "better living through chemistry," the critics of capitalism shifted their own vocabulary and concerns to the language of risk and the risk of environmental catastrophe, and placed science in the role of regulator of capitalism.

It is striking that Philip Kitcher's *Science, Truth, and Democracy* (2001) retraces the path of the 1930s Leftists discussed here (Turner 2003). The notion of planning science for the public good, the problem of authority over selecting that good and of its dependence on scientific advance, and even the idea of science as a map useful for multiple purposes, which are the key elements of the text, are already there in the 1930s literature, as is the idea of the contract between science and society and the slogan that experts should be on tap rather than on top. Kitcher's updating of these arguments runs into a difficulty that Comte and Pearson recognized: the social sciences do not, at least yet, provide the right kind of knowledge. Kitcher's concern is with the problem of using the social sciences to predict the outcomes of decisions to implement or develop particular forms of science and technology, and he concedes that they are not capable of providing the scientific guidance needed.

With the idea of a Rawlsian standard for evaluation in mind, as well as the new emphasis on risk, he proposes something more modest: that we take adequate knowledge of effects as an ideal standard by which to judge decisions. He treats the problem of authority in terms reminiscent of Saint-Simon's labeling of workers as "associates," by suggesting that decisions be made on the model of the family, by which he means an association in which each protects the interests of the other, in a paternalistic way, if necessary. He acknowledges the disparity in knowledge between participants. He assumes, as Saint-Simon did, that in a relation of family-like association, the conditions for trusting acceptance of superior knowledge will hold, and those with knowledge will not abuse their power out of self-interest or self-delusion, that they will be, in Bernal's phrase, "communists." But he does not explain how scientists will acquire the character needed to exercise this self-control, and he cannot explain how any of this might work, any more than Saint-Simon could explain how politics would be replaced the administration of things. Instead, he opts to treat these unworkable models as ideal moral standards for evaluating what is actually done, thus avoiding the difficult issues that Bernal at least attempted to address.

The problems with which Condorcet began, and with which Saint-Simon grappled, are still the problems of a reasonable Left. To achieve the goals of human betterment requires the greater use of science, and also for people to become more scientifically educated or at least to become deferential to science. This program did not succeed, as the literature attempting to improve on it implicitly acknowledges. So, what a Left conception of science needs today is a way to replace it, to replace the goals it attempted to achieve, or to replace the attempted accommodation to liberal democracy. The critics of the Left, the remaining advocates for the Liberal conception of science, are not well-situated either. The Liberal conception of science depended, by its own account, on institutions that guaranteed the independence of scientists in their choice of projects and on an ethos or community tradition that was strong enough to constrain its members, and on a kind of science not concerned with consensus but with refutation. As Jarvie pointed out, it underplayed the dependence of science on the wider society, indeed, was in denial about it (2001, p. 559). But it is also true that the idealized model of science they constructed is farther than ever from the reality of science today. As the "replication crisis" has made painfully evident, the constraints of the tradition and its former high valuation of refutation are now compromised or threatened. In part, this is a result of the success of science itself, which has transformed the former "community" of science in innumerable ways, from its grant-driven institutional structure to its extension into domains of policy and regulation which demand consensus rather than contention. Jarvie's message of 2001 is still relevant: we have not developed a solution to the problem of the relation of science and democracy. And the larger failure lies with the liberal conception of science, which, for the reasons Jarvie sets out, proved unable to address these changes.

References

Alexopoulos, Golfo. 2017. *Illness and Inhumanity in Stalin's Gulag*. New Haven, CT: Yale University Press.

Bernal, John D. 1939. *The Social Function of Science*. Cambridge: Cambridge University Press.

Buckle, Henry Thomas. 1924. *History of Civilization in England*. 2nd ed. New York and London: D. Appleton and Company.

Bukharin, Nikolai Ivanovich. 1969 [1921]. *Historical Materialism: A System of Sociology*. Ann Arbor, MI: The University of Michigan Press. Accessed April 10, 2017. https://www.marxists.org/archive/bukharin/works/1921/histmat/intro.htm.

Comte, Auguste. 1864. *Cours de philosophie positive [1830–42]*. Paris: Bachelier.

Conant, James B. 1947. *On Understanding Science: An Historical Approach*. New Haven, CT: Yale University Press.

Condorcet, Marie Jean Antoine Nicolas Cariat, marquis de. 1976 [1793]. Fragment on the New Atlantis, or Combined Efforts of the Human Species for the Advancement of Science. In *Condorcet: Selected Writings*, ed. Keith M. Baker, 283–300. Indianapolis, IN: The Bobbs-Merrill Co.

Hogben, Lancelot. 1957. *Statistical Theory: The Relationship of Probability, Credibility and Error*. London: George Allen & Unwin Ltd.

Huxley, Julian. 1932. *A Scientist among the Soviets*. New York: Harper & Bothers Publishers.

Jarvie, Ian C. 2001. Science in a Democratic Republic. *Philosophy of Science* 68 (4): 545–564.

Kitcher, Phillip. 2001. *Science, Truth and Democracy*. Oxford: Oxford University Press.

Mill, John Stuart. 1978 [1859]. *On Liberty*. Indianapolis, IN: Hackett Publishing Co.

Pearson, Karl. 1888. *The Ethic of Free Thought: A Selection of Essays and Lectures*. London: T. Fisher Unwin.
———. 1905. *National Life: From the Standpoint of Science*. London: Adam and Charles Black.
———. 1937 [1892]. *The Grammar of Science*. London: J. M. Dent & Sons.
Polanyi, Michael. 1951. *The Logic of Liberty: Reflections and Rejoinders*. Chicago: The University of Chicago Press.
Porter, Theodore. 2004. *Karl Pearson: The Scientific Life in a Statistical Age*. Princeton: Princeton University Press.
Stark, Johannes. 1938. The Pragmatic and the Dogmatic Spirit in Physics. *Nature* 1 (141): 770–772. First published as 'Weisse Juden' in der Wissenschaft. *Das Schwarze Korps*, 15 July 1937.
Turner, Stephen. 2003. The Third Science War. *Social Studies of Science* 33 (4): 581–611.

CHAPTER 21

Popper's Conception of Scientific Discovery and Its Relation to the Community of Science

H. T. Wilson

The study of science and scientific activity as an archetypal institution of rationality in the West was the beneficiary of at least three general approaches during the twentieth century. There was first the emphasis on discovery, truth, method, and cognitive standards in research and theory. Second was a more recent focus on the scientific community as a 'social system' whose professional and occupational characteristics needed to be studied as a specific set of socialization processes and as structures of reciprocity and communication. Third was a view of science as a set of activities whose structures of authority were characterized by tradition and dogma and which could be comprehended fully only if earlier, more idealized, views of science and scientific activity were understood to constitute ideology.

Apart from occasional references to the second concern with the sociology of science as it developed in the twentieth-century US social and intellectual setting, inspired largely by the work of Talcott Parsons (1951), my focus in what follows will be almost exclusively with the first approach.

H. T. Wilson (✉)
Professor of Public Policy & Public Law, Program Evaluation & Policy Analysis, York University, Toronto, ON, Canada
e-mail: htwilson@osgoode.yorku.ca

R. Sassower, N. Laor (eds.), *The Impact of Critical Rationalism*,
https://doi.org/10.1007/978-3-319-90826-7_21

The second concern, let it be noted, comported admirably with Popper's view of what the social sciences should ideally address their research and intellectual concerns to by confining itself for the most part to incremental studies of extant scientific practices within the emerging social system. Throughout the analysis, not surprisingly in light of his background influence on Popper, the work of Max Weber (1978, pp. 3–62) will loom large, particularly some of his key sociological concepts and his view of social research as a practice committed to reconciling an individualizing method with a generalizing objective (Cahnman 1964).

The first approach belongs pre-eminently to Popper and his followers, while the third is found most prominently in the work of Thomas Kuhn (1970a). Although Popper may perhaps be more popular with social intellectuals who esteemed the physical and biological sciences as distant models against which social research should be compared, often unfavorably, Kuhn's appeal was to social scientists and social theorists who took a relatively hostile attitude toward science's alleged superiority (Wolin 1968). What will become evident in what follows, however, is that those who favored Popper's analytical approach and interpretation had little warrant for doing so if they based it on his alleged support for either positivism, narrowly understood, or the unified science movement. Indeed, he was already well known for his groundbreaking and innovative arguments against both of these positions in *The Logic of Scientific Discovery* (1959) and elsewhere (Popper 1963).

This led Popper to resist being labelled a 'positivist' in the strongest possible terms, particularly where the effect of this label was to place him in the same camp as the logical positivist school of philosophy. It was their mistaken notion of what was involved in scientific activity itself, to wit, that its method and goal was to verify, or verify in principle, rather than to falsify statements and claims about reality, which from the beginning set his thinking apart from theirs (Ayer 1936). A further point put him completely in conflict with the unified science position, namely, his view that only the physical and biological sciences were justified in seeking 'truth' in their theorizing efforts, while the social sciences and allied disciplines were to be confined to a focus on 'success' in the interest of minimizing, and ideally eliminating, the alleged pernicious effects that invariably flowed from their commitment to truth rather than success (Popper 1945, 1957, 1972, pp. 262ff).

Popper's view of scientific activity appears to take its social and communitarian features largely for granted. Rather than making this inter-subjectivity the basic problematic in his work, he wanted to move beyond language without, however, foreclosing the possibility that communication may often be a source of confusion in research and related scientific activity. Popper feared that the study of science, no less than scientific activity itself, may be led astray by an overly reflexive approach and focus (1957, pp. 135–6, 142, 148–9). There are aspects of his views on this matter which bear a considerable resemblance to Bacon, even though Popper is clearly a Cartesian deductivist rather than a Baconian inductivist, for reasons that become apparent in *The Logic of Scientific Discovery.* According to Popper, no such practice or method of discovery exists, either in science or anywhere else (1959, pp. 27–48ff).

One result of submerging, but not by any means ignoring, the social and communitarian features of science is to underscore this activity as one carried out by individuals, albeit individuals of a particular kind. This is evident in the conceptual and methodological individualism which (following Weber) is so fundamental to Popper's philosophy of science. At first glance, he seems to view the community itself as an abstraction whose social character is exhausted in and through its individual members. Yet his emphasis on the cognitive aspects of scientific discovery cannot be reduced to a 'psychology of research' as Kuhn claimed. Indeed, the core principle of methodological individualism, as Weber pointed out and as Popper also believes, is that it combines an individualizing method with a generalizing objective in all forms of research (see Cahnman 1964; Kuhn 1970b).

Popper, it turns out, is a rationalist in much the same way that Weber is a rationalist. The influence of Kant, and the continental tradition of science and philosophy, cannot be overestimated in a study of his views. Far from being either a 'mechanist' or an 'emanationist', Popper insists that communities or groups of any kind are made up of individuals whose sum total is a different kind of whole from those whose close proximity is purely accidental and who lack a common bond. Like Weber, Popper insists that the nature of social reality is *constitutive* rather than constituted. Indeed, we may want to go further to suggest that for both Popper and Weber, there is often a clear sense of social reality as *constituting*, that is, 'in process', and therefore far less capable of being subsumed under disputes about whether the whole is or is not more than and/or different from the sum of its parts.

In the event, Weber's nominalist bias in favor of 'ideal types' as the most fruitful theoretical constructs to employ in social research aimed at combining an individualistic method with a generalizing objective is clearly definitive for Popper, since 'the task of social theory is to construct and to analyze our sociological models carefully in descriptive or nominalist terms, that is to say, in terms of individuals, of their attitudes, expectations, relations, etc.—a postulate which may be called "methodological individualism"' (1957, p. 136). Thus Popper is no opponent of the view that the natural and social sciences share a common approach. He rather argues that the general agreement on method which guides both enterprises needs to be understood as typological and rationalistic, and that the efforts in each case should be different because of the nature of the subject matter involved.

In neither case is induction valid, or even feasible, as a method. Instead, scientific method must be understood for what it is, 'deductive, hypothetical, selective by way of falsification, etc.' (Ibid., p. 137). Holistic and historicist theories of philosophy and social science are incorrect and potentially (if not actually) dangerous because their search for social 'laws' presumes the validity of a (false) positivistic conception of discovery in the natural sciences as an experience resulting from the process of induction and the commitment to verifiability rather than falsifiability. It is their utopian engineering or technological orientation which explains their wrongheaded view of what science is and what scientists do. Once they understand what is actually involved in these activities, they will no longer be able to employ 'science' as an exemplar of and justification for holistic and historicist approaches to social inquiry (Ibid., pp. 67–93; Popper 1963, pp. 336–346, 355–363; also Marcuse 1973).

The result of what Popper does when he studies what scientists do and what science is is thus *social* theories about scientific activity, where the general method is similar but the direction is not. Relevant here is his position on the two intellectual traditions so central to modern Western science: the empiricist and the rationalist. We have already mentioned Popper's preference for Cartesian rationalism and his criticism of the Baconian theory of induction, which he claimed to have 'demolished' in *The Logic of Scientific Discovery* (1959, pp. 27–48) and elsewhere (Popper 1972, pp. 1–105). His concern was to show that the idea that induction as a method could actually explain the growth of knowledge and scientific progress generally was demonstrably false. Human beings, scientists among them, do not constitute a passive *tabula rasa* on which experience imprints itself, first in the form of specific impressions, then in the form of generalizations formulated on the basis of these impressions.

Human rationality is far more active and dynamic than this. As a rationalist, Popper wants to expose this limited view of humans as rational animals by arguing that we reason *rationally* rather than mainly as a matter of habit, as David Hume and British Empiricism generally claimed (Popper 1963, pp. 3–30, 50–59, 120–135). This is particularly the case with scientists engaged in scientific activity, but obtains to a lesser extent, depending on the circumstances, for any human activity which requires thought and thinking. As he shows in his later work on this subject, Popper views human beings as active, world building, rationally reasoning animals whose highest achievement is modern Western science, alongside institutions and practices of critical rationalism in philosophy necessarily based on a clear demarcation between sense, nonsense, and dangerous thoughts.

Thus Popper, though an enemy of the 'common sense theory of knowledge' based on induction from specific observations to generalizations about them, is a proponent of 'common sense rationality' (Weber 1978; Cahnman 1964). Here the individual is understood to be a rational animal whose concern and interest in solving particular problems is what brings him to intellectual and scientific activity in the first place. Science is, for all investigative purposes, an individual enterprise carried out by human beings whose capabilities for science are an instance of, and a development out of, their character as substantively rational animals. They are not attempting, or rather should not be attempting, to verify particular hypotheses but to disconfirm or refute them. This is not simply an academic distinction, for an emphasis on verification is not really science, properly understood, at all. (The best we can hope for is corroboration (Popper 1959, pp. 251–281).)

It falls short of science because it is usually easier to verify or find confirming instances than it is to falsify, particularly given the human tendency to close the process and bring it to an end prematurely. This emphasis on falsification thus constitutes the highest standard to which the scientist should aspire, regardless of the assumptions that are made when using the 'best-tested' theory. Verification may constitute, or come to constitute, the scientist's proximate assumption, but he must be ready and willing to return to this higher standard once it intervenes and demands of him that he make the proper choice (Ibid., pp. 215–250). This higher standard requires the scientist to subject the best-tested theory to merciless criticism, criticism which takes the form of relentless efforts at disconfirmation and refutation, with no quarter given to any attempts to foreclose this process prematurely, a process that, by all rights, can, should, and must never come to an end.

But this practice as a standard can only be the case if we are speaking of groups and communities of individuals, not individuals per se. In effect, Popper's criticism, and apparent refutation, of the theory of induction and his support for conjectures and refutations generally, coupled with utilization of the best-tested theory, are not fully consistent with one another. There appears to be an inconsistency between the way he treats scientific discovery as an individual cognitive enterprise and the way he deals with science's progressive development as an institution. Hume's problem was whether past observations justify future predictions, and Popper argues that they do not, while the scientists he tells us about often must necessarily assume, like the rest of us, that they do. Hume's point was that this preference for the best-tested theory was not rational, even though it might be defended rationally, but could only be explained by reference to custom and tradition (Popper 1963, pp. 120–135).

Thus critical rationalism and the 'open society' it allegedly promotes when working at its very best does not allow us to assume that we reason rationally, since to do so we would have to return to some 'presuppositionless' (*standpunktslosigkeit*) point every time, and no such point exists. Without this non-rational aspect, in other words, science as institution and process cannot be explained and made sense of. Rationalism alone is insufficient, though eminently acceptable as an attitude, preference, or ideal. The question is not whether there is a specific scientific tradition characterized by criticism. Popper has argued that there indeed is, and has made a strong case that the practice of critical rationalism constitutes a tradition or habit at the center of science as an activity and institution. The issue instead is whether science is sufficiently intermingled with everyday life that its practice has no choice but to rely on assumptions which not only have a non-scientific but a non-rational basis as well.

On the other side of the issue is another difficulty, more obvious to those interested in the sociology of science. It too concerns the status of science as a rational activity, that is, the role of reason in scientific activity rather than the rationality of theories about scientific discovery. Popper's ideal of openness could conceivably negate itself and actually work against 'progress', since it is just possible that an individual may be mistaken in believing he is mistaken. This assumes for the moment something which is, of course, untenable, namely, that the way science as an institution has actually developed historically, and the 'growth of knowledge' as a whole, is the only possible way it could have developed. For this would mean that the method or procedure of trial and error, while a necessary condition for explaining the scientific developments we have before us, could not possibly be a sufficient condition. In effect, deterministic thinking on the matter would be transformed into a claim of inevitability, effectively negating the function of this method or procedure from the other side, as it were.

Indeed, we might be warranted in arguing against the *rationality value* of any hypothesis that those who initially supported it were not willing to stick with in the face of apparent refutations and disconfirmations. Here we would be thinking of some balance between 'loyalty' and a totally critical attitude which could well serve to make every conjecture, however well thought out and based on publicly repeatable tests and evidence, too provisional for it to be propounded in the first place. (Parsons emphasizes the obligation of scientists to tentatively accept hypotheses, and thereby suspend disbelief, once tests have corroborated them [1951, pp. 353–354].) Popper admits that this view may be an idealized overestimate of what we can realistically expect from working scientists at more than one point in his work. In doing so he appears to claim that falsifiability may in fact be valid primarily as a *collective or corporate* procedure rather than an individual one. After all, the claim that human beings are substantively rational animals cannot gainsay the fact that they are also social in nature and that this sociability is no less fundamental than their substantive rationality.

Thus it appears to be the case that, substantively rational though human beings are, it is the *public character* of the knowledge resulting from this collective or corporate procedure that is the key to explaining science's growth (Ziman 1968). Here, following Popper, we may be warranted in substituting 'evolution' for growth, for a process of intellectual 'selection' does seem to be operating in the effort to make sense of the growth of knowledge. However, this is not meant to imply that the collectivity corrects the errors of individual scientists, however tenacious, for there may be no collectivity as such, but rather a plurality of sub-groups organized around particular problems and hypotheses. (This anticipates part of Kuhn's criticism of Popper in Kuhn 1969 and elsewhere.) Nevertheless, the tenacity of individuals may well be the crucial element which makes it possible for a particular hypothesis to eventually come into its own in the face of initial hostility or skepticism. Without such dogged determination, theories having greater 'truth value' might never be able to weather the storm which invariably arises when they challenge established paradigms.

Popper puts the matter clearly and concisely in the following description of such dogged determination:

> The dogmatic attitude of sticking to a theory as long as possible is of considerable significance. Without it we could never find out what is in a theory—we should give the theory up before we had a real opportunity of finding out its strength; and in consequence no theory would ever be able to play its role in bringing order into the world, of preparing us for future events, of drawing our attention to events we would otherwise never observe. (1963, p. 312, n. 1)

Now we may be able to discern two important limits to Popper's philosophy of science as it pertains to the logic of scientific discovery. Both challenge his rationalistic bias and suggest the central role of custom, convention, and tradition in the development of an institution and process which could not be understood to have 'progressed' as it has without giving at least some pride of place to these decidedly non-rational and collective characteristics.

While one might argue that in making this claim we are only challenging the capacity of social theories to 'capture' the essential character of the activity under scrutiny, Popper does seem to be speaking to an ideal of science rather than to what most individual scientists do in their everyday work life. But this is precisely the point: Popper's claim is that science develops, grows, and therefore 'progresses' less by what most scientists do most of the time than by what exemplary scientists uniquely do very rarely. Popper seems less interested in the everyday life aspect of science, whatever he claims, at least when compared to those who are mainly interested in the sociology of science as a collective institution characterized by discrete rules of secondary socialization. His is not a 'great man' theory of discovery, invention, or innovation by any means, but his approach nevertheless does conform to the investigative tenets of Weberian methodological individualism in its determination not to be trapped in the zero-sum game of the individual versus the collective when it comes to the matter of how discovery occurs in science.

Indeed, I am strongly tempted to argue that for Popper, the concept and practice of scientific discovery as such comprehends more mundane achievements as well as the remit claimed by 'great man' discoveries, which occur only rarely and by a very select few. This is completely compatible with methodological individualism and provides Popper's theory of scientific discovery with the flesh and bones that is only possible for a social theory which is sensibly social. While at first glance, it appears that Popper may be overselling us on the adventure that is science, even given his intended focus on scientific discovery, such openness is understandable in light of his view that there really is no logic of scientific discovery premised on induction from specific observations to generalizations about them (1959, pp. 27–48, 78–111). Nevertheless, it is clearly premature to argue that only the very rarest, earth-shaking discoveries merit this description. Indeed, this overlooks Popper's individualism and forgets his emphasis on human beings as substantively rational animals, and science as one of our most impressive achievements.

Perhaps it would be more accurate to accuse him of making more of the day-to-day professional and intellectual work that Kuhn refers to as 'normal science' *because* of his individualistic orientation and bias. Seen in this light, scientific discovery can be understood to be a more everyday aspect of scientific activity *as well as* a uniquely distinctive feature of it for Popper. This 'deflation' thus has the effect of making it more 'available' to members of the community than conventional definitions of discovery would imply. By making discovery an ordinary as well as an extraordinary feature of science, Popper shifts the emphasis away from what the 'community' may think constitutes enduring contributions to the growth of its knowledge and toward what the individual scientist subjectively believes. There is clearly an element of 'gratification' present in such a conception, as if Popper were trying to answer Weber's claim in 'Science as a Vocation' that science's belief in its own progress is supremely irrational. Popper is responding by insisting that such progress, recast in terms of the individual's own sense of values and achievements, is therefore clearly substantively rational in Weber's sense of the term (1946, pp. 138–140).

Popper's emphasis on conjecturing and refuting, however complemented by a kind of collective confidence on the part of the scientific community and the role of the non-rational in the growth of knowledge generally, is nevertheless alleged to be a search for the *truth* about empirical reality. His ideal takes the form of an updated version of Socratic discourse, updated in the sense that all of the dialectical and reflexive features of theorizing allegedly central to Plato have been factored out in favor of Alfred Tarski's view of objective truth as the correspondence between statements about the world and the facts of the specific claim at issue. The status of these facts, in turn, depends on how successfully the theories they provide a case for are able to stand up to determined and unceasing efforts to falsify them (Popper 1963, pp. 114–119, 215–250). Having said this, the ideal of truth so central to Popper's view of the scientific quest can only be approximated, achieved incrementally at best, and only tentatively established. His faith must then reside in his unwavering belief that the truth about empirical reality is in fact being ever more closely approximated over time, not necessarily in a straight line, but that it is nevertheless a progressive development, if not a process in its own right.

What really serves to emphasize the commitment such a faith entails is the scientist's obligation to believe in this progressive development without having recourse to material 'realizations' in the form of technological manifestations subject to criteria of 'success' as evidence for the veracity of this belief. Popper is firmly against the instrumentalist view of the role of theories in science which argues that theories ought to be regarded mainly as tools in research, and judged strictly in terms of their differential utility or functional capacity to this end. He believes that such a focus is appropriate only to technological rules, where such a concern with success is not only warranted but essential. With Mario Bunge (1966), he views the long-term effects of this confidence in an instrumentalist view of theories in science as potentially and actually injurious to science, thereby underscoring his commitment to truth as the goal of theories in science. Popper made his position on the matter clear in the following passage contrasting pure and applied science, the first aiming at truth and the second at success. It was an important statement in the emergence of what was to become Popper's 'evolutionary' theory of the growth of knowledge:

> Admittedly, the growth of applied knowledge is very similar to the growth of tools and other instruments: there are always more and more different and specialized applications. But pure knowledge (or 'fundamental research' as it is sometimes called) grows in a very different way. It grows almost in the opposite direction to this increasing specialization and differentiation. As Herbert Spencer noticed, it is largely dominated by a tendency towards increasing integration and towards unified theories. (1972, p. 262)

Clearly this emphasis on 'truth' puts Popper closer to the view of science as natural philosophy than most of his contemporaries, and appears to locate it somewhere between philosophy and technology. The objective, after all, is an ever closer approximation to the truth, albeit about the physical realm of nature rather than the Platonic world of Forms or Pure Being, against which physical reality is viewed as the world of 'mere appearance'. Indeed, Popper contrasts Socrates and Plato on this score, arguing that only Socrates is a true critical rationalist, while Plato's is at best a 'pseudo-rationalism'. At the same time, Popper similarly resists breaking down any of the walls which serve to separate science from technology. Critical rationalism provides a basis not only for falsification aimed at disconfirmation of theories allegedly in good standing with the scientific community, but for organized skepticism because a highly structured and interdependent set of first principles oriented to proper conduct exist, and are accepted implicitly as prior 'givens' in scientific research and discovery. In this regard, it is not so much the existence of prior agreement, as the character these first principles impart to science, understood as a particular corpus of knowledge constitutive of a normative order, which is so significant.

Science is therefore something of a constitutional republic when it is functioning as it should, but what scientists do is not always what science is for Popper. Members of the scientific community have emerged from a secondary socialization process which equates the satisfaction of certain meritocratic performance norms not only with good conduct but with rationality itself. The screening which education and training have accomplished initially, and which work and effort in some specialty should only reinforce, given science's character as mainly 'public' knowledge, is the point of departure for all efforts at openness, but openness only to individuals who have also accepted this environing framework, science's 'rule of law' or constitution understood both as opportunity and as limit. These terms could hardly be more appropriate, not only because of the common loyalties and first principles of science and liberalism but because of Popper's belief that the scientific community and scientific knowledge provide an ideal, albeit a distant ideal, against which the incremental practice of 'piecemeal social engineering' for both political practice and social research should measure themselves. This, in spite of the different and unique direction taken by science, whose search for truth, in contrast to political practice and social research, is alleged to be integrative, rather than differentiating.

There is no space to discuss in any detail the two major controversies that addressed Popper's views on scientific discovery and its relation to the community of science. One concerns Kuhn's criticism of Popper's view for being too prescriptive, idealistic, and unrealistic. Kuhn claims that scientific work is rarely, if ever, based on falsification and the fervent attempt to disconfirm theories presently in good standing, but is mainly an ordinary activity which he refers to as 'normal science'. Only in critical periods, when a prevailing 'paradigm' is under sustained attack, can a version of the falsification/disconfirmation model be seen to hold, and then for only a relatively brief time, before the scientific community begins its return to normalcy, acknowledging either the reinstatement of the earlier paradigm or the ascendancy of its successor. Realizing how rarely it turned out that actual scientific revolutions were acknowledged to have occurred under his definition of 'revolution', Kuhn later qualified his earlier rigid distinction between normal and extraordinary science, admitting that his own definition of revolution was too exclusive to capture what in fact did constitute revolutionary developments in science. This had the effect of narrowing the distance between the ordinary or 'normal' and the extraordinary significantly (Kuhn 1963, 1970b; see also Toulmin 1972).

A second controversy has also already been alluded to, and was mainly addressed to Popper's view that the social sciences should limit their theoretical endeavors to incremental efforts to improve existing practices and institutions, what he called 'piecemeal social engineering'. While it is true that Popper did believe that holistic and historicist theories not only could but actually already had had tyrannical results in and through their instantiation in practice, his argument against them had a logical basis as well. In the preface to *The Poverty of Historicism*, he claimed that since the course of history is strongly influenced by the growth of knowledge and this growth of knowledge cannot be predicted, therefore neither can the future course of history be predicted. It is for this reason that 'we must reject the possibility of a theoretical history; that is to say, a historical social science that would correspond to theoretical physics. There can be no scientific theory of historical development serving as a basis for historical prediction' (Popper 1959, pp. vi–vii). My own work (Wilson 1991) and that of others (Marcuse 1973) disputed Popper's assumption that holistic and historicist theories convert to tyrannical practice this automatically, if at all, and also stressed the contrast between deterministic and inevitabilistic theories. In the event, Marxian and critical theory were admittedly deterministic, but were unequivocally anti-inevitabilistic in their animus.

This idea of what amounts to a logical relation between theory and practice, such that the proposing of a 'holistic' or 'historicist' theory is tantamount to its instantiation and endorsement in practice, is even alleged to work both ways for those who believe it. Thus, the formal end of the Soviet Union between 1989 and 1991 was touted as conclusive evidence that Marx's theories of analysis and development had been invalidated as well, thereby confirming what Popper and his supporters had always claimed. A more careful look at Marx's theory of analysis and development shows clearly how problematic the Soviet case was for anyone claiming that it provided a *de facto* example of his theory in practice, since it definitely did *not* adhere to the Marxian developmental paradigm. Most fundamentally, no substantial period of sustained capitalism had preceded what Joseph Stalin, after 1936, would claim was its subsequent 'socialist' and 'communist' stages of development. This claim was undermined further by the realization that the pre-modern class structure of the Soviet Union lacked the *sine qua non* of capitalist development itself, namely, a dominant bourgeoisie, with its tributary sub-elements, all or most suffering developmental 'crises', and facing a deprived, hostile, and soon-to-be revolutionary proletariat. Even Vladimir Lenin (1932), in prescient anticipation of what was to come, had felt it necessary to drastically modify the Marxian analysis in order to accommodate the clearly 'deviant' case of Czarist Russia.

Popper's already impressive corpus of work was increasingly perceived to be even more cohesive analytically than had been the case earlier in his ongoing researches, as he moved nearer to the end of his life and began to garner the recognition and esteem that had for too long been denied him. It could now be said with justice that the result was a full-fledged philosophy in its own right, as opposed to a philosophy of science and/or social science alone. As a philosophy, it is coherent yet open to review and ongoing disputation, in some cases, because of subsequent research and discoveries that were not yet known at the time when he was working out his thoughts, in others, because social theories, by Popper's own admission, are often, for better or worse, prey to the vagaries of fad and fashion. In my opinion, what provides Popper's massive body of work with its most secure ballast, however, is the anchorage one discovers upon a careful reading of his analyses of the ancient philosophers, particularly the pre-Socratics. Popper came to the philosophy of science because in his view it constituted, alongside critical rationalism, as its governing procedure, the most impressive cognitive achievement in the history of humankind; yet much of the source of this achievement is to be discovered in his analyses of the pre-Socratic philosophers.

This is why he refused to yield to positions which would allow any remit short of the search for truth as the goal not only of science but of the growth of human knowledge as a whole. You can see this clearly in his discussions of Parmenides and Heraclitus, for example, among many others among the pre-Socratic group of philosophers. As a social theory of science in its relation to philosophy as a whole, and the growth of knowledge so central to its progress, in whatever this consists, Popper's philosophy is eminently successful in its sustained sponsorship of truth, based on its commitment to criticism, as it is operationalized in attempts at falsification aimed at disconfirmation of reigning as well as emerging theories. Whether this is claimed to stand as an ideal or as actual practice seems less important than that it exists and regularly can be shown to be a standard or goal of inquiry in the service of the growth of knowledge. If some of us think that truth and success also permit of theoretical procedures and methods of analysis other than his, it can and must be seen as part of the growth of knowledge as well. This is given in the fact that it is reality that ultimately determines the fate of all theories, including those in science.

References

Ayer, Alfred J. 1936. *Language, Truth and Logic*. London: Victor Gollancz.

Bunge, Mario. 1966. Technology as Applied Science. *Technology and Culture* 7: 329–347.

Cahnman, Werner. 1964. Max Weber and the Methodological Controversy. In *Sociology and History*, ed. Werner Cahnman and Alan Boskoff. New York: Free Press.

Kuhn, Thomas S. 1963. The Function of Dogma in Scientific Research. In *Scientific Change*, ed. A.C. Crombie, 347–369. London: Heinemann.

———. 1970a [1962]. *The Structure of Scientific Revolutions*. Chicago: University of Chicago Press.

———. 1970b. Logic of Discovery or Psychology of Research? In *Criticism and the Growth of Knowledge*, ed. Imre Lakatos and Alan Musgrave. Cambridge: Cambridge University Press.

Lenin, Vladimir I. 1932 [1916]. *State and Revolution*. New York: International Publishers.

Marcuse, Herbert. 1973. Karl Popper and the Problem of Historical Laws. In *Studies in Critical Philosophy*, 191–209. Boston: Beacon Press.

Parsons, Talcott. 1951. *The Social System*. New York: Macmillan.

Popper, Karl R. 1945. *The Open Society and Its Enemies*. London: Routledge and Kegan Paul.

———. 1957. *The Poverty of Historicism*. London: Routledge and Kegan Paul.

———. 1959 [1934]. *The Logic of Scientific Discovery*. London: Hutchinson.

———. 1963. *Conjectures and Refutations*. London: Routledge and Kegan Paul.

———. 1972. *Objective Knowledge*. London: Oxford University Press.

Toulmin, Stephen. 1972. *Human Understanding*. Princeton: Princeton University Press.

Weber, Max. 1946. Science as a Vocation. In *From Max Weber: Essays in Sociology*, ed. Hans Gerth and C. Wright Mills, 129–156. New York: Oxford University Press.

———. 1978. *Economics and Society, Volume I*. Edited by Guenther Roth and Claus Wittich. Berkeley: University of California Press.

Wilson, H.T. 1991. *Marx's Critical/Dialectical Procedure*. London: Routledge.

Wolin, Sheldon. 1968. Paradigms and Political Theories. In *Politics and Experience*, ed. Preston King and B.C. Parekh. Cambridge: Cambridge University Press.

Ziman, John. 1968. *Public Knowledge*. Cambridge: Cambridge University Press.

PART IV

Film Studies/Aesthetics

CHAPTER 22

Some Thoughts on Artists' Statements

Jeanette Bicknell

Introduction

I used to be employed to teach philosophy at a university-level institution for art and design. One of the many pleasures of doing so was that I got to see a lot of art. The art on display, whether it was figurative drawing and painting, large-scale installation pieces, intricate metalwork by materials design students, or conceptual art, was almost always accompanied by artists' statements; more often than not, I took the time to read them. Some genuinely illuminated the works they accompanied and enriched my experience of them; some were autobiographically frank and expressed the artist's joy of creation; others made reference to theories and insights gained from the student's compulsory humanities courses; and some came over (for one reason or another) as a bit confused.

It seems to me that artists' statements can perform two different functions and often perform both. First, an artist's statement allows the artist to provide information to viewers that is not necessarily discernible from the work. Such information might include the artist's own attitudes toward the work and thoughts about it or information on the materials used and on specific techniques. Second, an artist's statement can contextualize a work. It can direct the viewer to see, interpret, or appreciate a

J. Bicknell (✉)
Independent Scholar, Toronto, ON, Canada

R. Sassower, N. Laor (eds.), *The Impact of Critical Rationalism*,
https://doi.org/10.1007/978-3-319-90826-7_22

work in specific ways. Of course, an artist's statement cannot *compel* viewers to have a particular experience of a work; the most it can do is to suggest or guide viewers in a certain direction.

Artists have long written about their work, but this writing was often personal (in diaries and letters) or intended for other artists and students. Artists with strong theoretical or political interests have written articles, manifestos, and treatises. However, artists' statements, and their public display alongside a work or a body of work, are a comparatively recent phenomenon. They have become ubiquitous only in the last 20 years or so. Artists' statements are now standardly required as part of the application process for many art world institutions, including graduate programs, residences, jobs, grants, and so on. According to a 2008 survey of art colleges, 90 percent of the respondents said that their institution taught the writing of artist statements (Garrett-Petts and Nash 2008). All students in the art program in the university where I taught were required to write one by their fourth year. In fact, the expectation that an artist will provide a statement has become so entrenched that when I asked a senior administrator what would happen if a new artist simply refused to do so, he looked genuinely perplexed and said that he "couldn't imagine" such a situation.

This institutional entrenchment of artists' statements makes it all the more remarkable that they have received little philosophical scrutiny. I have modest ambitions in this chapter. I want to open up the discussion by pointing out a few areas where our approach to artists' statements will have a bearing on deeper philosophical questions. I owe the approach I will take to the position set out by Ian Jarvie's "The Objectivity of Criticism of the Arts" (Jarvie 1967). In this article, Jarvie attacks the view that judgments on works of art are irremediably subjective and proposes a Humean alternative. Objectivity is to be found not in individual critics or their pronouncements but in the traditions of rational arguments about the arts and in the institutions which carry on these traditions. In seeking objectivity in the arts, we seek minimally biased institutions with built-in mechanisms for self-criticism. Objectivity is "a sort of democracy of opinion and criticism, within which the truth may be pursued" (Ibid., p. 3). Critical claims which enjoy wide agreement and have stood the test of time are more likely to be true than those that have not, although we must always be alert to the possibility that even widely endorsed claims may be upended one day.

Accordingly, my claims in this chapter are intended as provisional conjectures, robust enough to be criticized and improved upon, rather than as settled conclusions. It is too early for there to be a "received" view or settled consensus on the topics I discuss here. Philosophers of art have had so far little to say about the phenomenon of artists' statements. The closest they have come, it seems to me, is to discuss the philosophical significance of artworks' titles (Fisher 1984; Gombrich 1991; Levinson 1990). Titles and artists' statements raise some similar problems. Both prompt questions about the ontology of works (are they parts of works or not?) and interpretation (how much weight should be given to them?).

However, the similar problems raised by titles and artists' statements should not make us overlook important differences. Throughout the history of art, there have been many different conventions regarding the entitling of artworks, and philosophers over-simplify these at their own peril. The practice of entitling started to take hold in the early modern period, with the growth of the art market in the Netherlands. Titles were as likely to be chosen by art dealers as they were by artists themselves (Yeazell 2015). Artists' statements, in contrast, are presented as being composed by artists to accompany their works. (Although this may not be actually the case, as there are professional writers who will craft artists' statements on behalf of artists for a fee.) Furthermore, artists' statements are supposed to be genuinely informative, which is not necessarily true of titles, even of titles chosen by artists themselves. A title might only minimally designate a genre or subject ("Madonna and Child," "Still life") or place a work in a sequence ("Abstract no. 7"). Artists' statements demand more, on the part of artists, critics, and viewers alike.

Let me now turn to three areas that seem particularly fruitful for thinking about artists' statements: ontology, interpretation, and evaluation.

Ontology

On one common view, some singular artworks are concrete objects. (I set aside conceptual works because they raise special difficulties.) In contrast to this view, a contextualist would argue that that the work's identifying properties depend on certain features of the context in which it was created (Levinson 1990). On a contextualist view, would an artist's statement about a work, say, a painting or a sculpture, constitute part of that work? If so, then many works which seem to be in one medium are actually hybrid works, consisting of painting and text or found objects and text, and so on. Furthermore, many of the artworks of the past, which traditionally were not accompanied by artists' statements, are rather different things than contemporary works in similar media that do include artists' statements. The ontological "advantage" (an explanatory written text that is part of a work's essence) enjoyed by contemporary works will certainly have implications for how we interpret them.

The alternative to the view that an artist's statement is part of a work would hold that artists' statements are more like tools to help us understand artworks, which are themselves concrete objects. Although this view seems plausible, we might want to consider whether statements written by a work's creator are different from explanatory statements written by others, such as curators, historians, collectors, and art critics. In practice, artists' statements would seem to have a different ontological status and demand from us a different kind of respect. (Whether they *should* have a different status is of course a separate question.) If it seems plausible that artists' statements are not on a par with, say, curator's statements, then we might want to consider how far their "different status" extends. Does it extend to their being constitutive of the works they elucidate? Imagine a sculpture displayed without the written statement that the artist meant to accompany it. Is the work incomplete, or does it merely lack a useful tool that might have helped us to appreciate and interpret it? Can there be a general rule about this, or do we have to decide on a case-by-case basis? If the position that artists' statements are constitutive of works is correct, then these are hybrid works.

Matters are not any clearer if we venture beyond the position that artworks are concrete objects. According to a rival account of the ontology of art, artworks are processes, action types, or performances (Davies 2004). On such views, viewers attend to a "focus of appreciation" rather than a work, strictly speaking. For example, the painting by Diego Velazquez called "Las Meninas" (1656) is the focus of appreciation of the artwork that is in turn constituted by all the actions committed by Velazquez while painting that particular canvas. "Las Meninas" didn't come with an artist's statement, of course. Viewers have only the canvas to appreciate. If they chose, they can also research the history of the painting, learn about the context of its creation, and rival theories about the painting's meaning. But there is no definitive statement by Velazquez imparting his views on the work.

Compare "Las Meninas" with a contemporary painting that is intended to be accompanied by an artist's statement. It would seem clear that the statement is part of the artist's actions or performance and so is part of the work. So, on a "process" view of the ontology of art, artists' statements are constitutive of the work. But then, so are all the actions taken by the artist in creating the work. Does the statement have any special status? Does it constitute part of the focus of appreciation? We have solved the ontological question (the contemporary painting is a process encompassing both a concrete object and a text), but the implications that follow for the work's appreciation are not clear.

Amie Thomasson (2005) has argued that the answers to ontological questions about art are to be found in our practices, more concretely, in the practices of those whose words and actions ground the terms. On her view, claims about the ontology of art are not mind-independent facts that we may "discover" the way we discover facts about the natural world. Rather, the competent users of these terms ground the ontology of such terms through their usage. In the case of painting, competent "users" would presumably be artists themselves, art teachers, critics, and consumers.

Setting aside the adequacy of this view, I think that artists' statements pose a particularly tricky problem because there are as yet few established practices. (And how long does a practice have to persist before it is considered "established"?) Some people make a point of ignoring artists' statements, while others read the artist's statement before looking at the work. (In fact, if there is a crowd around a work and I can't get to it, I may read the artist's statement while waiting my turn to view the work.) Another way of weighing the ontological status of artworks with accompanying statements is the fact that students learn to write statements during the course of learning effective use of materials, color, and design, as if the written words are as important to craft as the artwork itself. In other words, the writing of artists' statements is integrated into artistic pedagogy and therefore supports the view that artists' statements are part and parcel of any work of art.

Stephen Davies (2003) has suggested (sensibly, I think) that our ontology should be revealing of how and why art is created and appreciated. By this he means to say that philosophy should not stray too far from artistic practice and appreciation. Whichever ontology of art we start with, the status of artists' statements will have to be confronted. While I lean toward the view that artists' statements are constitutive of works, I think that some philosophical work needs to be done here before this view could be robustly defended.

Interpretation

How much interpretive weight should we give to artists' statements? If statements are constitutive of works, then their status must be acknowledged and taken seriously when discerning the meaning of those works. I have already said that we seem to treat artists' statements differently than similar writings by curators, collectors, and critics. Since an artist's statement is "from the horse's mouth" (so to speak), we accord it a kind of deference that we don't necessarily extend to writings by people who are not the creators of that particular work.

Is this practice defensible? It would seem perverse to ignore artists' statements in our interpretation of works. Artists often take great care in crafting these statements, and even if we decide that they are secondary to the art object, they are at least one source of information about the works they accompany. At the same time, we expect that audiences will bring their own background and experience to the works they view, and that multiple interpretations are reasonable (if not all equally plausible). We don't assume that the composer of a musical work will necessarily give the single correct or even the best performative interpretation of that work. In literary criticism, the "intentional fallacy" is said to be committed when an interpreter assumes that the meaning intended by the author of a work is definitive. Furthermore, artists (like any of us) may be self-deceived. Their sincerely expressed beliefs about their own work may be mistaken, and placing too much weight on an artist's statement might impede coming to a different, more plausible interpretation of the artwork.

A further complication arises: Must we assume that all artists' statements are transparently written? What of statements that are ironic, or even deliberately misleading? For example, photographer David Leventi writes, of his series *Bjoerling's Larynx: World Famous Opera Houses* (2012): "I have photographed each house systematically from the spot at center stage where a performer would stand." But looking at these photographs, it is obvious that some of them were taken not from center stage but from the back of the house facing toward the stage. An assistant at the gallery where I first saw the photographs told me that Leventi was not always allowed access to the stage, and that is why some of the photographs are taken from the back of the house. What are we to make of the contradiction between the statement and the work? Is it a simple oversight? That is possible, but seems unlikely as the rest of the statement has obviously been written with care. Does it imply an injunction to see the work in a certain way? Or are we perhaps meant to think about the gap between ideal aspiration and actual realization?

Can an artwork carry a meaning that has not been indicated or sanctioned by the artist in his or her statement? We do not need to imagine an extreme case, where a statement claims that a work that endorses X, while an interpreter believes that the work condemns X. I am thinking of a case where an interpreter sees some meaning in a work that an artist has not specifically indicated and is perhaps not even aware of, but might endorse. This question is especially interesting with regard to young artists. Their work may have historical echoes and resonances that of which they are not fully aware (because they have been poorly informed or poorly trained in the history of their craft, and perhaps because of both). Yet the artist might well endorse such significations if they were pointed out, whether by a mentor, gallery owner, curator, viewer, or potential buyer who likes the piece exactly for these significations. In such cases, writings by a more knowledgeable critic or curator might be more illuminating than the artist's statement.

How much interpretive weight to give artists' statements is further complicated by the fact that artists, like the rest of us, sometimes change their minds. It is not difficult to imagine an artist in her 40s who decides with the wisdom of hindsight that she was mistaken in important ways about the works she made while in her 20s. The more mature artist may have a very different relationship to her work and a very different understanding of it than did her earlier self. If given the opportunity, she may wish to revise or modify her earlier statements about these works. These considerations further undermine any reliance on artists' statements in the interpretation of art.

Critical Evaluation

From comparing films among friends to reading critical notices about the latest show or gallery exhibition, much of our engagement with artworks consists in criticism, evaluation, and relative ranking of them. We compare artworks with one another, discussing the supposed flaws of one against the merits of another. We discuss an artist's mature output in light of his or her earlier work. We devise and debate "top ten" lists and argue about who really deserves that prestigious prize. In evaluating an artwork and defending that evaluation to others, how much regard (if any) should we give to artists' statements? Does the answer change if we are considering not just a single artwork, but a body of work or even an entire career?

If an artist's statement is ontologically constitutive of a work, it would seem to follow that critical evaluation of that work should include a consideration of the artist's statement. If a work of art is (nothing more nor less than) a concrete object, and the artist's statement is not constitutive of it, then it would seem that we should decide on a case-by-case basis what weight (if any) to place on it.

Whether we think of artist's statements as part of a work or as a source of information about a work, the role we give them in critical evaluation may turn on the importance we place on artists' intentions. I've heard art students speak as though the single most important evaluative criterion was whether or not a work fulfills the maker's intentions. For those holding such a position, the artist's statement is a crucial source, as it may be the only direct and sanctioned evidence we have of the artist's intentions for the work. Can a work be successful even if it does not fulfill the artist's intentions as manifest in the corresponding statement? Such a situation certainly seems possible; one can esteem a work for many different reasons, not all of which have to do with the maker's intentions. Yet any critic making such claims would be in a tricky position: Praising the art while diminishing the significance of the artist's own hopes for it might be seen as patronizing.

There is another, even less comfortable possibility: The artist's intentions may be banal. We may conclude that the statement is woolly and reflects a sketchy understanding of fashionable intellectual trends. In such a case, should we allow our negative reactions to the artist's statement color our judgment of the art object? Or, should we try to set aside our response to the statement so as to appreciate the value of the work on its own? I was faced with such a dilemma when viewing the photographs by Leventi mentioned earlier. Now, his artist's statement for the series is neither woolly nor intellectually shallow. On the contrary, it is well-written and for the most part illuminating. But Leventi claims that his photographs of empty theaters "freeze for eternity the instant before a performance takes place" (2012). I found this claim to be disconcerting, at the very least. Performances do not take place in empty theaters. I would go so far as to argue that nothing taking place in an empty theater counts as a performance. So, I'm not sure that I even understand what Leventi means here. While these misgivings marred my enjoyment of the photographs, would it be proper to let them affect my evaluation of them? A critic who shared my view might come to different evaluations of the photographs whether she judged them "as photographs" or "as photographs that purport to capture something important about performance."

Conclusion

The institutional entrenchment of artists' statements completes a transition in the way we regard artists. The artist is no longer seen as a craftsperson who is mentored or apprenticed to a master. Instead, contemporary artists are formally educated or credentialed individuals who are supposed to display their education in words. I suspect that there is another reason why viewers pay attention to artists' statements. Because these statements are purported to be a direct communication between artist and viewer, they provide a substitution for (or the impression of) a genuine personal connection. Clearly, artists' statements would benefit from and repay increased philosophical attention.

Acknowledgments I am grateful to Ian Jarvie, Raphael Sassower, and the audience at the Canadian Philosophical Association 2012 annual meeting held at Wilfrid Laurier University for comments and discussion.

References

Davies, David. 2004. *Art as Performance*. Malden, MA: Blackwell Publishing.

Davies, Stephen. 2003. Ontology of Art. In *The Oxford Handbook of Aesthetics*, ed. J. Levinson, 155–180. Oxford: Oxford University Press.

Fisher, John. 1984. Entitling. *Critical Inquiry* 11 (2): 286–298.

Garrett-Petts, Will F., and Rachel Nash. 2008. Re-visioning the Visual: Making Artistic Inquiry Visible. *Rhizomes* 18. Special issue on "Imaging Place." Accessed August 1, 2017. http://www.rhizomes.net/issue18/garrett/index.html.

Gombrich, Ernest H. 1991. Image and Word in Twentieth-Century Art. In *Topics of Our Time: Twentieth-Century Issues in Learning and in Art*, ed. E.H. Gombrich, 162–187. London: Phaidon.

Jarvie, Ian C. 1967. The Objectivity of Criticism of the Arts. *Ratio* 9: 67–83.

Leventi, David. 2012. Artist Statement. Accessed February 23, 2011. http://www.davidleventi.com/#/artist%20statement/.

Levinson, Jerrold. 1990. Titles. In *Music, Art, and Metaphysics: Essays in Philosophical Aesthetics*, ed. Jerrold Levinson, 159–178. Ithaca: Cornell University Press.

Thomasson, Amie L. 2005. The Ontology of Art and Knowledge in Aesthetics. *Journal of Aesthetics and Art Criticism* 63: 221–229.

Yeazell, Ruth B. 2015. *Picture Titles: How and Why Western Paintings Acquired Their Names*. Princeton: Princeton University Press.

CHAPTER 23

Confusing the Scientific and Moral Appeals of Suppressing Vice

Augustine Brannigan

Introduction

It is not difficult to determine when fears about the adverse effects of mass media became a concern of civil society and motivated attempts to censor the worrisome elements in popular fiction under obscenity law. This coincided with the changes in mass media in the twentieth century: the rise of radio, TV, movies, and the Internet. However, history shows that sexuality had been a much more acknowledged and accepted element in pre-European social life before the rise of Christianity. In his history of pornography, Walter Kendrick (1988) records how the eighteenth-century excavation of the ruins of Pompeii, buried in 79 AD by the eruption of Mt. Vesuvius, brought to light household statues and paintings from ancient Roman culture that revealed an extraordinary sexual frankness. Residential frescoes displayed nude females, couples making love, satyrs having sex with goats, and phalluses in relief in paving stones and on the walls of houses. Brothels were apparently common, and the activities of women of pleasure were captured graphically on interior walls of people's homes. Such vivid sexual representation was found so offensive during

A. Brannigan (✉)
Department of Sociology, University of Calgary, Calgary, AB, Canada
e-mail: branniga@ucalgary.ca

R. Sassower, N. Laor (eds.), *The Impact of Critical Rationalism*,
https://doi.org/10.1007/978-3-319-90826-7_23

excavation that the materials were housed in a secret museum in Naples. The museum catalogue was considered X-rated. The "pre-pornographic" culture began to unwind in the nineteenth century as traditional sexual inhibitions came under pressure with the spread of literacy. Librarians objected to the publication of Mark Twain's *Adventures of Huckleberry Finn*, and the book was banned from the Concord Library in Massachusetts as "trash suitable only for the ghetto" (Twain 1885). Toronto's *Public Health Journal* targeted the stories of Jesse James, Deadeye Dick, and Nick Carter in adolescent magazines in the 1920s as filth purveying media requiring a legal ban from circulation in the mail. In the late 1930s, the appearance of comic books was greeted by alarm, and all common law jurisdictions attempted to ban them under obscenity laws. The criminalization of crime comics remains in Canadian obscenity law to this day. The provocative covers, the adult crime themes, and the celebration of violence were said to promote juvenile delinquency, racism, and homosexuality, according to psychiatrist, Fredric Wertham, whose book, *Seduction of the Innocent* (1954), and whose articles in popular magazines struck a chord with the public. Here was a psychiatric authority who shared the public's misgivings about the influence of perverse literature. According to Wertham, "we do not maintain that comic books automatically cause delinquency in every child reader … But we found that comic-book reading was a distinct influencing factor in the case of every single delinquent or disturbed child we studied" (in Crist 1948). Every one of them?

An American inquiry into the causes of delinquency in the early 1950s led by Senator Kefauver raised questions about the influence of mass media on youthful misconduct. Questions of imitation of criminal behavior had been raised earlier in the Payne Foundation studies of motion pictures in the 1920s–1930s. Paul Lazarsfeld studied the influence of radio on political opinions in the 1940s–1950s and found little evidence for significant shifts in political opinion related to campaign speeches (Hovland 1959). US academic funding of mass-media effects, especially television, was in its infancy until the work of Albert Bandura (1973) raised questions about the vulnerability of youth to messages of violence in children's programming, including popular Saturday morning cartoons. Bandura advanced a model of influence known as "social learning theory" in which viewers could acquire new behaviors through vicarious experience, such as seeing other people benefiting from a specific behavior and mimicking it with expectations of gaining a similar benefit. Monkey see; monkey do. The 1960s saw the emergence of lab studies of imitative violence as well as long-term field studies of the correlation between violent media exposure and aggressive behavior. Interest in the question of media effects grew as the crime picture became increasingly grim throughout the twentieth century.

In 1972 the US Surgeon General issued a report on the effects of television violence that cautiously expressed some concerns about the inadvertent effect of violent entertainment on viewers. Its conclusions were couched in highly conditional language: Television "can" induce short-term mimicry in children, and it "can" instigate an increase in aggressive acts; however, the evidence "does not warrant the conclusion that television violence has a uniformly adverse effect nor the conclusion that it has an adverse effect on the majority of children" (Liebert and Sprafkin 1988, p. 113). At the instigation of mass-media researchers, a new report was prepared a decade later under the auspices of the National Institute of Mental Health: *Television and Behavior: Ten Years of Scientific Progress and Implications for the Eighties*. The 1982 report updated the empirical findings and concluded that the evidence supported inferences of a causal link between exposure to violent TV and aggressive behavior among viewers, suggesting that there was a wide consensus among social scientists about this fact and that the conclusions were based on the "convergence" of different kinds of evidence, none decisive on its own, but convincing when taken together.

In Canada, the 1985 Fraser Commission on Pornography and Prostitution was less convinced about the effect of pornography on sexually aggressive behavior (Casavant and Robertson 2007). Berl Kutchinsky's studies of the de-regulation of pornography in Denmark were associated with a decline in sex crime (1991), a conclusion re-iterated by Milton Diamond and Ayako Uchiyama (1999) in their studies of Japan and Hawaii. Fast forward to the second decade of the twenty-first century: Internet pornography is rife; anyone with a browser can access an unprecedented amount of free, sexually diverse, and often amateur, sexually explicit material. This has not escaped the notice of professionals: The American College of Pediatricians (2016) has reported deleterious effects on children from exposure to such materials and its effect on their tendency toward aggression, a vivid echo of historical concerns about the unanticipated effects of exposure to sex and violence in popular entertainment. How are such changes in aggression purportedly explained?

Social Learning Theory

In the twentieth century, psychologists advanced two different theories of learning based on completely different mechanisms. First, "classical conditioning" was demonstrated famously by Pavlov's experiments with dogs. Dogs, like other mammals, salivate when they sense the presence of food. Pavlov produced salivation by pairing an unconditioned stimulus (a bell) with a conditioned stimulus (food) and showed that the animals learned to salivate at the sound of the bell unrelated to the presence of food. Hence psychologists inferred that human feelings of, say, joy and revulsion could be created by social structures that conditioned perceptions and feelings with what were *arbitrarily* paired with other sensations. Boy likes girl. Girl wears perfume. Boy subsequently is stimulated by the perfume.

Second, there is operant conditioning based on the work of Watson and Skinner with rats. The differential rewards given to the rats for choosing alternative T-mazes and other pathways accelerate the animal responses, and they become adept at problem solving and learning new behaviors. This was also called trial and error learning.

In the early 1960s, Bandura sought out a third explanation of learned behavior peculiar to humans and reflective of their sophisticated cognitive abilities (Bandura, Ross and Ross 1963). Bandura's social learning theory was the chief theoretical basis for lab studies of the behavioral and attitudinal changes attributed to the electronic media. It had some peculiarities that ultimately would impede its credibility. The first was the remarkable narrowness in what it examined. It was preoccupied with aggression, and it focused on such popular programs as *Batman* and the *Road Runner* cartoons as though aggression was the main "lesson" of such programs and as though this was the only effect worth noting. Aggression became a proxy for all that was problematic with youthful behavior. Nobody seemed to wonder whether movies that portrayed bravery, altruism, or self-sacrifice encouraged viewers to become ennobled, caring, and sensitive. The virtually exclusive focus on "aggression" proved remarkably one-dimensional. This went hand in hand with a presupposition about the peculiar vulnerability of children. By implication, social learning occurs primarily in childhood. The famous longitudinal studies of the lagged effect of early childhood TV viewing habits (Huesmann et al. 1984) seem to assume an age-graded model in which violent media exposure created persistent antisocial behaviors.

When one sees the research agenda of the media effects experts in the 1960s–1980s, it is hard to escape the proposition that the effects of greatest interest were decided a priori in the world of respectable fears about the vulnerability of children and women and the apparent invasiveness of the new technologies. For every study on the positive effects of children's programming on literacy, there were 25 studies on the consequences of sex and violence.

Is Social Learning Theory Valid?

Social learning theory appears to be quite modest when removed from specific technologies (TV) and specific prohibited behaviors (aggression), and when conceived in more general terms. There is nothing in the theory establishing that people are peculiarly vulnerable to TV as opposed to other sources of information, parents, siblings, peers, schoolteachers, newspapers, books, and neighbors. Social learning theory cut its teeth on television and the subsequent video technology without reference to how it worked in all human history prior to the intrusion of national networks into family life. The very idea of social learning only emerged with the alarm bells associated with the rise of television. The intellectual study of social learning appeared with a specific social agenda: censoring children's programming and promoting social values in the name of science and "public health." The social agenda of regulating television was premised on the understanding that it had become one of the most important sources of learning, that it was the source of decisive social models, and hence one of the most important determinants of behavior. The experiments were designed to demonstrate this view. The early Bobo doll studies were designed to showcase how children learned bad habits from watching cartoons. The fact that the Bobo doll used in Bandura's experiments was designed to be punched and that some forms of physical roughhousing in the context of play can be wholesome behavior were overlooked.

There was a definite moral cast in the attitudes of psychologists to popular television dramas and cartoons in the 1960s. They assumed that the violence in *Batman*, *Superman*, and cowboy fiction was rewarded and that the viewers were consequentially converted to the Dark Side. Yet most "violent" fiction results in the punishment of the wicked and the reward of the virtuous (Fowles 1999). This is the cathartic attraction of virtually all popular fiction, a point consistently missed by those who believe that catharsis in fiction is only interesting if it alters behavior afterward and somehow lets the viewer "blow off steam." The catharsis I refer to is the identification with the hero, the suspense experienced as the hero faces danger, our fear and contempt of the villains, our anxieties along plot lines, and our relief as the Calvary rides in to restore justice. Without catharsis, fiction would be lacking dramatic attraction. Why would we presuppose that human viewers would miss the almost clichéd moral nature of popular entertainment, the Punch and Judy portrayal of good and evil? In my view, the failure to grasp this point has been a defect in social learning theory and has led researchers to overlook the situational impact of short-term media exposure on emotional volatility in the lab, something that is frequently misinterpreted as imitation. I will return to this point later.

What are the actual mechanisms that contribute to social learning? Psychologists write at times as though changes in behavior occur below the threshold of consciousness. When social learning occurs, are we simply talking about absorbing cultural images as though they were normal and unnoticed? In other words, we grow up in the Shire and think as hobbits. Or we grow up in Rivendell and think as elves. Though worldviews differ, they are experienced as natural, like growing up Catholic or Protestant in Ulster. But the concept of selective exposure to certain worldviews is not a specific mechanism of learning, unless we acquire *conditioned* responses to categories of people in the same process, for example, Catholics are lazy, Protestants are smart. This conceptualization of social learning becomes classical conditioning. In social learning theory, male consumers of pornographic fiction, for example, come to react to women purely based on their capacity to cause arousal as though this was not already endemic in "real life."

A second issue is whether the observed behavior that is mimicked results in a reward that we vicariously approve of and pursue for our own benefit. We mimic another's behavior because we see that it resulted in certain benefits. Is this subconscious? Or would we be aware of consequences of action and consciously make similar choices? Is that not a case of "judgment," choosing wisely through generalization from the stimulus to everyday life? This conceptualization of social learning becomes operant conditioning which is equivalent to "reasoning" or "problem solving." Following this line of inference, social learning is not a distinctive form of learning at all. The argument from social learning theory reduces to the observation that we take our own culture for granted (selective exposure leading to classical conditioning) and/or we maximize our utilities as classical economics suggests. There is nothing mysterious here. Perhaps the focus on children makes such selective exposure, on the one hand, and self-reflection, on the other, less than obvious since we assume children are unreflective. This indicates that social learning is premised on the inability of children to reason. The extension to the pornography effects on adults is ill advised, unless the adults influenced by pornography are basically acting like children, something that seems counter-intuitive.

Is there another approach to social learning that tacitly acknowledges the power of fiction to create arousal? When we observe that children may be acting more immaturely or impulsively in the aftermath of provocative images, do we perceive the cathartic, stimulating, or just plain entertaining effects of drama? When we "act out" in the aftermath of exposure to a "violent" cartoon, have we somehow been seduced by a mechanism of which we are only dimly aware? Is there some magical effect that sweeps us up into activities that would be terminated with mature self-reflection and/or social control? I think that something like that does occur and that it is tied to catharsis, that is, arousal.

Let me explain my reason for this line of thinking. If social learning theory explained aggression, there would be one academic discipline with a vested interest in determining whether it was valid or not: criminology. The methods of psychologists and criminologists tend to be quite different. The case made for the effects of media on aggression follows the protocols associated with the Buss paradigm. In the Buss aggression paradigm, subjects are put into a high state of emotional discharge, typically by being insulted by a confederate. Subjects are then assigned to a treatment condition (e.g., one of several different kinds of film), and then in a supposedly unrelated phase of the experiment are asked to administer shocks to the person who earlier had insulted them. More recently the proxy for aggression is administering the provocateur high levels of "hot sauce" (Busching et al. 2016). The experiment picks up the media effect of the intervening treatment usually only in the presence of the high state of arousal. Leonard Berkowitz puts it this way: "the observer will exhibit the highest aggressive reactions if he is emotionally aroused at the time, believes his aggressive actions will have favorable rather than unfavorable consequences, and thinks the observed victims had deserved the injury inflicted on them" (1971, p. 18). So, the evidence of harm is somewhat oblique. As in the Milgram paradigm, subjects are encouraged to administer shocks as legitimate punishments, and then their behavior is equated with illegitimate aggression. However, we now know that in the Milgram paradigm, perception of harm to the subjects leads to lower levels of aggression (Brannigan 2013, pp. 11–12).

The limited utility of the lab evidence for the impact of media on aggression is suggested by the treatment of this subject in the classic textbooks in criminology. James Wilson and Richard Herrnstein (1985, p. 337ff) review the case for television and the mass-media impact on crime in *Crime and Human Nature*. Of the field studies, they say: "the best studies come to contradictory conclusions, and even when all doubts are resolved in favor of a causal effect, they account for only 'trivial proportions' of individual differences in aggression" (Ibid., p. 346). When experts assembled by the National Research Council reviewed the 1982 NIMH report, they concluded that televised violence "may" be related to aggression, "but the magnitude of the relationship is small, and the meaning of aggression is unclear" (Wilson and Herrnstein 1985, p. 353). "Even giving to existing research the most generous interpretation, viewing televised violence cannot explain more than a very small proportion of the variation in aggressive acts among young persons" (Ibid., p. 353). Similarly, Ray Kaplan and Richard Singer (1976) argued that "this research has failed to demonstrate that TV appreciably affects aggression in our daily lives." Seymour Feshbach and Richard Singer reported no evidence from field studies that violent fantasy programming aroused aggression. In fact, there was some evidence "that exposure to aggressive content on television seems to reduce or control the expression of aggression in aggressive boys from low socioeconomic backgrounds" (1971, p. 145). In Jonathan Freedman's review of the field studies, he reported that "not one study produced strong consistent results, and most produced a substantial number of negative findings" (1988, p. 158). David Gauntlett reported "the search for 'direct' effects of television on behavior is over: Every effort has been made and they simply cannot be found" (1995, p. 120).

The issue of media effects is reviewed in a leading criminology monograph, *A General Theory of Crime*, by Michael Gottfredson and Travis Hirschi (1990, p. 67ff). They point out that television viewing at age eight predicts "theft, motor-vehicle accidents, trivial nonviolent offending, drug consumption, and employment instability, behaviors hard to attribute to the number of shootings or fistfights watched on television twenty years previously" (p. 69). Similarly, Leonard Eron reports that childhood aggression predicted "social failure, psychopathology, aggression and low educational and occupational success" 22 years later (1987, p. 440). Why?

All kinds of dysfunctional behaviors tend to cluster in the same persons. Individuals who are "aggressive" do not tend to specialize in aggressiveness but exhibit a short temporal horizon or impulsiveness. "It therefore seems unlikely that the specific content of television programming viewed at age eight could contribute independently to subsequent levels of aggression" (Ibid.). According to Gottfredson and Hirschi, "it is all an expression of the same stable underlying traits of impulsivity that made provocative entertainment attractive in the first instance" (1990, p. 67).

The Magic in Arousal Theory

Earlier I alluded to psychological mechanisms that might supplant social learning explanations. We return to that issue here. In *Seductions of Crime*, Jack Katz argues that to explain the attractions of deviance, "one must believe in magic." His is a phenomenological approach to deviant behavior that emphasizes the foreground of experience in the commission of crimes. Where most criminologists stress the causal role of background factors such as family conflict, community disorganization, and class conflict, Katz argues that the clue to explaining crime is the "sensory attractions of doing evil," the physical pleasures of sneaky thrills, righteous slaughter, and stickup. Katz refers to "magic" as the process by which people let themselves be seduced by the criminal project: "To believe that a person can suddenly feel propelled to crime without any independently verifiable change in his background, it seems we must almost believe in magic" (1988, p. 4). As a phenomenologist, Katz rejects causal explanations, and suggests that the key to behavior is the powerful effect of emotions. In Katz's perspective, people let themselves be seduced by their emotions, surrendering to circumstances and situations, although he characterizes this as "an artifice" in the sense that this process involves an element of self-delusion. The ego is capable of resisting temptation but "gives in" as though compelled by emotions. For example, a person who encounters humiliation can re-fashion it as an act of self-defensive rage, and strike out at the provocateur in "righteous" indignation. The impulse to kill probably accompanies this emotional arousal but is successfully stifled by self-control.

Reviewing social learning theory, I suggested that there might be a non-obvious process operational in media exposure that is neither equivalent to classical nor operant conditioning. I would now add that it does not involve mimicry at all but appears to be an artifact of emotional arousal. The typical evidence for media effects in both TV and pornography studies discovers aggressive effects only when subjects are already highly aroused (usually by being insulted by a female confederate of the experimenter) and only when the experimenter requires that the angry subjects administer shocks (or hot sauce) to their female aggressor under the pretext of a learning exercise. William Fisher and Guy Grenier (1994) attempted to replicate some of the critical work from Edward Donnerstein (1983) that showed how certain kinds of pornography facilitated aggression against women. Donnerstein reported that films that combined both aggression and erotic elements boosted shock levels given by male subjects to female targets compared to films containing either aggressive or erotic stimuli alone. Fisher and Grenier gave the subjects the option of skipping the shock experiment and proceeding directly to the de-briefing. Most subjects, even if angered, opted to skip this opportunity for aggression and proceed toward the de-briefing. In other words, the link between stimulus films and the aggressive outcome reported in earlier work was an artifact of the design that forcibly attached "aggression" to media exposure.

But the nature of the aggression is far from clear. In an earlier work, Donnerstein reported that "aggressive behavior in subjects who have previously been angered has been shown to be increased by exposure to arousing sources such as aggressive-erotic films, physical exercise, drugs, and noise" (1980, p. 279). Dolf Zillmann and Jennings Bryant (1984) used films of an eye operation and discovered that these boosted levels of aggression. Percy Tannenbaum (1972, p. 330) discovered that a humorous film had the same result. "Even viewing *Sesame Street* or *Mister Roger's Neighborhood* induced a threefold increase in aggression among preschoolers who initially measured low on aggressiveness" (Fowles 1999, p. 28). The underlying process as understood by Zillmann and Bryant is called "excitation transfer." The subjects are angry because they have been insulted. The anger response tends to abate with time, but before it does so, it becomes re-energized by an intervening arouser. Pornographic films have an impact not because they contain a message about women as social learning theories suggest but because they are highly arousing (see Zillmann 1984). Just as a nauseating film of an eye operation is emotionally arousing, so does an aversive noise or a humorous film. The initial retaliatory impulses that arise from provocation are boosted by the intervening stimulus, and the subjects react with more anger than they are probably aware of. But the effect is short-lived since once the anger has abated, the aggression is no longer fueled by the emotional distress at the heart of the Buss paradigm.

Also, in their long-term design, Zillmann and Bryant suggest that pornographic films lose their ability to arouse after repeated exposures, so the potential of films to promote retaliatory violence becomes self-limiting. Speaking of TV, Dolf Zillmann comes to the same conclusion. "It would thus appear likely that repeated exposure to dramatic portrayals of violent crime reduces rather than increases affective reactions" (1991, p. 123). This suggests a major conclusion. Rather than attributing media effects to social learning, the mechanism of excitation transfer and the emotional impact of exposure to media offer a more parsimonious explanation of the experimental effects; it is also consistent with the lack of long-term effects of media on behavior.

Competing Legal Doctrines in Pornography and Prostitution

One of the paradoxes of the "criminological turn" in the psychology of pornography is that it connected contemporary laws of pornography with prostitution. Canadian court judgments were characterized by a series of inconsistent legal perspectives following the explosion of widely circulated pornography in the 1980s–1990s (*Caligula*, *Deep Throat*, *Debbie Does Dallas*, etc.). The Canadian obscenity law criminalizes adult entertainment when it exceeds the Canadian "standards of tolerance" based on "the undue exploitation of sex, or sex in combination with crime, horror, cruelty or violence." Since the law is federal, such a standard is assumed to be uniform across the provinces. It is also assumed to be objective, not some judge's personal opinion ("pornography, I know it when I see it"). And increasingly its condemnation could not be grounded in a religious or moralistic perspective. The secular position came to a head in a landmark Calgary case, *R. v. Wagner* (1985). This was one of the first cases where the trial judge concluded that the establishment of "undue exploitation" was based on the harms of the materials in question had on their consumers. The experimental evidence of media effects was critical to the verdict, although it was based on the social learning theory of media effects.

The theory in *Wagner* and other cases was reviewed by the Supreme Court of Canada. Where the decision in *Wagner* accepted the harm criterion for "undue exploitation," *Butler* (1992) established that the evidential base for undue exploitation of sex could be met if the court considered the materials to be inherently degrading to women, promoting sexually callous attitudes, or so violent as to promote mimicry. Explicit sexual pornography was off the hook. Films, books, paintings, and personal materials could attract censorship, either as consumer products or as materials imported through customs, if they were deemed to be harmful. This was construed in two ways. Such materials degraded the viewers' attitudes to the persons depicted (women) and/or increased the likelihood of violence against women.

Ian C. Jarvie's analysis of degradation in pornography (1991) predated the *Butler* decision by a year. His conclusions were premised on an anthropological analysis of the alleged status of women depicted in pornography. It identified a decline in status due to pornography from a previous high water level. Because of pornography, women in everyday life are "smeared" by the life of Linda Lovelace, for example, because viewers assume her theatrical performances are natural, voluntary, and uncoerced; they lead viewers to assume all women behave similarly, thereby discounting the status of all women. But inconsistencies in Lovelace's various biographies about her voluntary-then-coerced-then-something-else performances failed to validate the claim of degradation. Jarvie's refinement of the illustration consists of a thought experiment that describes the case of libertines who engage voluntarily in all manner of sexuality, then decide to film their escapades, and eventually share them with friends. The charge of degradation fails since the films report what participants do voluntarily. There is no loss of status and hence the case for degradation fails. *Butler* challenges this reasoning by disposing of the anthropological view with the third-person harm perspective: Even innocent first-person viewers of the anthropological films may be "harmed" in the sense that their attitudes and sexual dispositions may take a turn for the worse. In other words, even voluntary conduct from everyday life could be objectionable because it is obscene (harm-causing), a term related to degradation partially (but meaningfully), insofar as it degrades the attitudes and behaviors of those exposed to it, irrespective of whether the subjects creating the materials are not coerced.

The problem for the rationalist is that scholars who accept the *Butler* criterion for obscenity will be perplexed by the law of prostitution. Here I would point out two things. Jarvie argues that (a) most pornographic performers are prostitutes, albeit of a higher pay grade compared to streetwalkers. "I have argued that historically and by definition the bulk of actresses in pornography are prostitutes" (Jarvie 1991, p. 21). He then claims (b) "prostitutes are no doubt a degraded class of women, even to themselves" (ibid.). My objective here is to expose the fact that the Canadian law crosses over from a moralistic perspective to a harm criterion; that is, it moves from traditional, moralistic obscenity laws to a more rational or evidence-based perspective (notwithstanding the conclusions drawn earlier about the scientific weaknesses of that literature).

Prosecuting the Harm in Domination

Terri-Lynn Bedford is a Canadian prostitute turned dominatrix whose long, legal struggles led to a legal precedent in the Supreme Court of Canada in 2013 (Bedford 2011). It effectively abolished the legal suppression of prostitution. Bedford's conflict with the law began when the premises she operated as a domina in Richmond Hill were raided by police in 1994. Evidence showed that Bedford and her associates operated a house for practicing domination, sadomasochism, cross-dressing, and other paraphilias with paying clients. However, evidence showed that they did not participate in the usual sex acts associated with prostitution (intercourse, fellatio, etc.). The house was equipped with a dungeon to constrain clients, a throne for the worship of the domina, a mock schoolroom for discipline, a mock medical examination room, as well as a small stage to display the costumes and role-playing of cross-dressers. The police seized over 700 items, including costumes, whips, wigs, feathers, handcuffs, physical constraints, and all the other tools of the trade required for role-playing. Despite irregularities in the arrest process and the excesses of search and seizure, as well as confirmation by undercover officers of unsuccessful attempts to solicit sexual acts, Bedford was convicted at trial (1998) on the charge of operating a common bawdy house, and her conviction was affirmed in the Ontario Court of Appeal in 2000. A bawdy house is a location employed habitually for prostitution. The written decision failed to identify which acts constituted prostitution according to Canadian law. Bedford was fined $3000.

In 2007 Bedford was joined by Valerie Scott and Amy Lebovitch in a charter challenge led by Alan Young in the Ontario Superior Court. The challenge focused on key features of Canadian law designed to suppress prostitution that were unconstitutional: (a) suppression of communication for the purposes of prostitution, (b) pimping and living off the avails of prostitution, and (c) keeping a common bawdy house. These laws, the court was told, endanger the lives of persons who work in the sex trade, force them to seek their own security with violent pimps, and conduct their work in a dangerous and clandestine manner. Prostitution per se has never been illegal in Canada. Justice Himel accepted the arguments and declared the laws unconstitutional and of no force (Bedford et al. v Canada 2010). These laws "are not in accord with the principles of fundamental justice and must be struck down. These laws, individually and together, force prostitutes to choose between their liberty interest and their right to security of the person as protected under the Canadian Charter of Rights and Freedoms." At the time of the charter challenge, evidence of the vulnerability of prostitution was made all the more compelling by the prosecution of Robert Pickton in Vancouver who was investigated in the disappearance of some 50 sex-trade workers, and convicted of murder in the case of 6 of them. He disposed of their remains in the food supply on his pig farm.

On appeal to the Supreme Court, the bench agreed with most of Himel's decision except the challenge to the communication law (Bedford et al. v Canada 2013). Because of nuisance, the court reasoned, society had a valid interest in suppressing public communication. However, for the most part, the Supreme Court removed the stigma associated with pimping, living off the avails, and operating a bawdy house. The lesson of the *Bedford et al.* case was that the *clients* were degraded, and they paid for it voluntarily. Those charged under the prostitution laws were the victims of justice, so the narrative of harm had shifted dramatically. The subtext of Jarvie's degradation analysis of pornographic actors is that the charge of degradation confuses a moral and an empirical question. His thought experiments undermined the claim that pornography was degrading by *implying* that it was a moral question. His rationale was that people who engage in voluntary sexual conduct in ordinary circumstance suffered no degradation.

Final Question: Have Morality and Harm Come Full Circle?

What was the final chapter in the Bedford saga? After the Supreme Court struck down key sections of the laws suppressing prostitution, it gave parliament a year to reflect on the de-criminalization of prostitution. In 2014 the government of Stephen Harper replaced Canada's legal approach with what has been dubbed "the Nordic model" (Skibrei and Holström 2014). This approach is based on the supposition that women are victimized by prostitution and that the remedy is to criminalize prostitution; the liability will be only on the buyers, not the sellers, on the assumption that they are unequal in status. It also removed liabilities for automatically defining a domicile as a bawdy house and those living therein. However, the law also forbids sex workers from advertising their trade. So, the Scandinavian approach falls short of validating or legitimizing prostitution per se, criminalizes the johns, and puts the workers back into the shadows in pursuit of their livelihood. One can only hope the appeal courts will re-discover the logic of *Bedford et al.*

References

American College of Pediatricians. 2016. The Impact of Media Use and Screen Time on Children, Adolescents and Families. https://www.acpeds.org/the-college-speaks/position-statements/parenting-issues/the-impact-of-media-use-and-screen-time-on-children-adolescents-and-families.

Bandura, Albert. 1973. *Aggression: A Social Learning Analysis*. Englewood Cliffs, NJ: Prentice Hall.

Bandura, Albert, D. Ross, and S.A. Ross. 1963. Imitation of Film-Mediated Aggressive Models. *Journal of Abnormal and Social Psychology* 66: 3–11.

Bedford, Terri-Jean. 2011. *Dominatrix on Trial*. Bloomington, IN: iUniverse, Inc.

Bedford et al v Canada (2010) ONSC 4264 (CanLII); http://canlii.ca/t/2cr62.

Bedford et al v Canada (2013) 3 S.C.R. 1101 2013 SCC 72 (CanLII); http://canlii.ca/t/g2f56.

Berkowitz, Leonard. 1971. Sex and Violence: We Can't Have It Both Ways. *Psychology Today* (December): 14–23.

Brannigan, Augustine. 2013. *Beyond the Banality of Evil: Genocide and Criminology*. Oxford, UK: Oxford University Press.

Busching, Robert, Johnie J. Allen, and Craig A. Anderson. 2016. Violent Media Content and Its Effects. *Oxford Research Encyclopedia of Communications*. Online. http://communication.oxfordre.com/view/10.1093/acrefore/9780190228613.001.0001/acrefore-9780190228613-e-1?rskey=ZKWxdR&result=1.

Casavant, Lyne, and James R. Robertson. 2007. *The Evolution of Pornography Law in Canada*. Ottawa: Library of Parliament. https://lop.parl.ca/content/lop/researchpublications/843-e.htm.

Crist, Judith. 1948. Horror in the Nursery. *Colliers Magazine*, pp. 95–97.

Diamond, Milton, and Ayako Uchiyama. 1999. Pornography, Rape and Sex Crimes in Japan. *International Journal of Law and Psychiatry* 22 (1): 1–22.

Donnerstein, Edward. 1980. Aggressive Erotica and Violence Against Women. *Journal of Personality and Social Psychology* 39: 269–277.

Donnerstein, Edward. 1983. Erotica and Human Aggression. In *Aggression: Theoretical and Empirical Reviews*, ed. R. Geen and E. Donnerstein, 53–81. New York: Academic Press.

Eron, Leonard. 1987. The Development of Aggressive Behavior from the Perspective of Developing Behaviorism. *American Psychologist* 42: 435–442.

Feshbach, Seymour, and Richard D. Singer. 1971. *Television and Aggression: An Experimental Field Study*. San Francisco: Jossey-Bass.

Fisher, William, and Guy Grenier. 1994. Violent pornography, Antiwoman Thoughts and Antiwoman Acts: In Search of Reliable Effects. *Journal of Sex Research* 31 (1): 23–38.

Fowles, Jib. 1999. *The Case for Television Violence*. London: Sage.

Freedman, Jonathan L. 1988. Television and Aggression: What the Research Shows. In *Television as a Social Issue*, ed. Stuart Oskamp. Newbury Park, CA: Sage.

Gauntlett, David. 1995. *Moving Experiences: Understanding Televisions Influences and Effects*. London: J. Libbey.

Gottredson, Michael R., and Travis Hirschi. 1990. *A General Theory of Crime*. Stanford: Stanford University Press.

Hovland, Carl. 1959. Reconciling Conflicting Results Resulting from Experimental and Survey Studies of Attitude Change. *American Psychologist* 14: 8–17.

Huesmann, L.R., L.D. Eron, M.M. Lefkowitz, and L.O. Walder. 1984. The Stability of Aggression Over Time and Generations. *Developmental Psychology* 20: 1120–1134.

Jarvie, Ian C. 1991. Pornography and/as Degradation. *International Journal of Law and Psychiatry* 14: 13–27.

Kaplan, Ray M., and Richard D. Singer. 1976. Television Violence and Viewer Aggression: A Re-Examination of the Evidence. *Journal of Criminal Psychopathology* 3: 112–137.

Katz, Jack. 1988. *Seductions of Crime: The Sensual and Moral Attractions of Doing Evil*. New York: Basic Books.

Kendrick, Walter. 1988. *The Secret Museum: Pornography in Modern Culture*. New York: Viking Penguin.

Kutchinsky, Berl. 1991. Pornography and Rape: Theory and Practice? *International Journal of Law and Psychiatry* 14: 47–64.

Liebert, Robert A., and Joyce Sprafkin. 1988. *The Early Window: The Effects of Television on Children and Youth, 3/e*. New York: Pergamon.

R v Butler (1992) 1 S.C.R 452. CanLII 124 (SCC); http://canlii.ca/t/1fsdj.

R v Wagner 1985 (AB QB) 1130 CanLII. http://canlii.ca/t/27t5b.

Skibrei, May-Len, and Charlotta Holström. 2014. The 'Nordic Model' of Prostitution is a Myth. *LSE Blog*, 3 January 2014, London: London School of Economics and

Political Science. http://blogs.lse.ac.uk/europpblog/2014/01/03/the-nordic-model-of-prostitution-law-is-a-myth/.

Tannenbaum, Percy. 1972. Studies in Film and Television Mediated Arousal and Aggression: A Progress Report. In *Television and Social Behavior, Vol. 5, Television's Effects: Further Explorations*, ed. George A. Comstock, Eli A. Rubinstein, and John P. Murray, 309–350. Washington, DC: USGPO.

Twain, Mark. 1885. https://www.pbs.org/wgbh/cultureshock/flashpoints/literature/huck.html.

Wertham, Frederic. 1954. *Seduction of the Innocent*. New York: Rinehart.

Wilson, James Q., and Richard J. Herrnstein. 1985. *Crime and Human Nature: The Definitive Study of the Causes of Crime*. New York: Simon and Schuster.

Zillmann, Dolf. 1984. *Connections between Sex and Aggression*. Hillsdale, NJ: Lawrence Erlbaum Associates.

———. 1991. Television Viewing and Physiological Arousal. In *Responding to the Screen: Reception and Reaction Processes*, ed. Jennings Bryant and Dolf Zillmann, 103–133. Hillsdale, NJ: Lawrence Erlbaum Associates.

Zillmann, Dolf, and Jennings W. Bryant. 1984. Effects of Massive Exposure to Pornography. In *Pornography and Sexual Aggression*, ed. Neil Malamuth and Edward Donnerstein, 115–138. Orlando, FL: Academic Press.

CHAPTER 24

World Three and Cognitivism: Philosophy in Film

Byron Kaldis

As Ian C. Jarvie argues: "As an abstract object the film seems to contain philosophy and we are able to think about it philosophically. Why do writers, directors, etc. put philosophy on to film? The answer is simple: they do not. Philosophy puts itself in their films."
—(Jarvie 1987a, p. 26)

Introduction

Jarvie's importance in philosophy rests as much on the kind of thinking he brought to bear on the subjects that interest him when he proceeds to analyze them in a critical and unprejudiced fashion as on his unique philosophical, often dialogical, style that combines erudition with a refreshing intellectual irony thanks to its no-nonsense disdain for fashionable superficialities. Although both this kind of thinking and the accompanying style are unique

B. Kaldis (✉)
Department of Philosophy, The National Technical University of Athens, Athens, Greece

Department of Philosophy, Moral Culture Research Institute, Hunan Normal University, Changsha, Hunan Province, China
e-mail: bkaldis@central.ntua.gr

R. Sassower, N. Laor (eds.), *The Impact of Critical Rationalism*,
https://doi.org/10.1007/978-3-319-90826-7_24

and therefore impossible to replicate, it is no doubt unfortunate, for philosophy and for the discourse in humanities generally, that such rare qualities are shunned in current philosophical practices. In a certain sense, Jarvie's texts should be taught as mush for their style as for their content and thus preserved in the imaginary museum of fine philosophical writing.

One of Jarvie's major contributions, being one of the longest serving academic journal editors, is the philosophy of the social sciences. Although this is one of my own areas of interest, this time I will discuss his well-known and influential work on the philosophy and sociology of film, even though to name only philosophy and sociology is not to do justice to the breadth of approaches he has brought to bear on film studies beyond these two fields. It is also a mark of his intellectual legacy that the various angles from which he has studied cinema as an art form combine in a unique methodological amalgam, bearing testimony to his distinctive philosophical style.

The present essay, *pace* the injunction just made regarding the all-encompassing viewpoint Jarvie adopts, will nevertheless concentrate on his philosophical views on film, assuming we could easily extract them from the richness of the whole without injuring its characteristic fullness. To undertake such an extraction, I propose to look at three of his texts where a single underlying thread runs through them: his book on the philosophy of film, his paper on the objectivity of criticism, and his paper on the rationality in creativity. The underlying thread in all three pieces is weaved into a major Popperian theme that I wish to explore subsequently in this chapter, bringing forth its significance for the philosophical analysis of film; this theme has to do with what Popper has called *objective knowledge*, his famous three-world ontology, and the items inhabiting his so-called World Three in particular. Though Jarvie only touches on Popper's World Three in relation to films, we can nevertheless be sure that his thought is guided in this respect by how film as an art form is to be understood in relation to World Three. I then go on to offer as a central thesis of this essay a particular reading of the Popperian worldview in light of Bernard Bolzano and Gottlob Frege as an alternative way of films being seen as engaged in philosophy.

The basic Jarviean tenets in this respect, congruent to World Three "objective knowledge" à la Popper, have to do with more general philosophical positions espousing anti-essentialism, anti-psychologism, anti-individualism/anti-subjectivism, anti-expressivism, and the metaphor of an evolutionary battle among theories faced with situations of problem-solving. All of this amounts to a critical-rationalist thought being embedded in public institutions of critical reason and tradition, as Jarvie repeatedly insists on, i.e. what can be called an *institutionalism of critical rationalism* underlying World Three abstracta and their objective significance.

Films and Philosophy

The debate over the relationship between philosophy and films has followed different strands. Offering a brushstroke taxonomy as a necessary starting point, we could inquire whether movies as special art forms can legitimately claim to engage in philosophy or "do philosophy via cinematic means in a mode unique to this medium." What "doing philosophy" amounts to or what the activity of philosophy is taken to be is a heavily contested issue. Recent discussions suggest that philosophy is a permanent struggle against opponents or a series of sophisticated objections and disagreements. Others argue that the meta-philosophical nature of the question of what philosophy is is itself what plagues the unavailability of a fixed answer and that there is no growth in philosophy compared to the growth of scientific knowledge. The Jarviean-Popperian notion of critical rationalism is compatible with the views about the nature of philosophy just mentioned. On the other flank of the taxonomy, there are issues associated with how philosophy looks at the arts generally and at the moving image in particular, that is, the philosophy *of* film. In the former case, which is the focus of this chapter, the issue is not merely whether films as such can be bearers of philosophical messages when paraphrased by a philosophically minded critic or viewer, but more contentiously, and thus for many scandalously, whether films can actually philosophize—contribute original philosophical theses not merely parasitic upon linguistic paraphrasing, i.e. in the same sense that bona fide, card-carrying texts of philosophy do.

In arguing pro and con the latter possibility, various philosophical positions are assumed. The effort to defend the possibility of films doing philosophy is connected to the specific art medium—the cinematic means—and whether it can yield the required results as texts normally do. Natural and artificial languages in which texts are weaved are normally considered as undisputed means of philosophical expression, but filmic pictures or moving images pose a problem when they are assimilated to texts given that their moving pictorial nature is far from being similar to their textual analogue. It is not just meanings or slogans or even theses of philosophy that may be conveyed by films that are at stake here; it is whether films can replace the argumentative and discursive power of language in enunciating philosophical propositions or even engage in philosophical criticism in the way we are accustomed to expect that written, technical philosophical texts normally do. When films are taken as doing philosophy, two claims could be made. The *weak claim* is that the content of a film can be interpreted by means of language as having a philosophical thesis; the *strong claim* (depending on how the cinematic means are dealt with) argues that the moving image on the screen is doing the philosophizing, not the philosopher-interpreter (see Wartenberg 2007 for a comprehensive treatment).

It is the claim of this chapter that there may be a parallel reading of the aforementioned Popperian three-world ontological economy (boosted by similar but not identical ideas voiced by Frege and even Bolzano) in regard to the possibility of objective knowledge without a knowing subject along the lines of *an objectively philosophical filmic content without a philosophical interpreter*. I should first clarify what is at stake when films are said to philosophize.

We may approach the issue in two ways. We may look at the films as igniting certain cognitive responses in the viewers that trigger in them philosophical thoughts. Call this the *cognitive approach* and the resulting reading of films as philosophy the *cognitive reading*, whereby the locus of philosophy is still the film itself which arouses in the minds of the viewers' cognitive effects (regardless whether they can subsequently be put into words). The second approach may be one that gives pride of place to the *ontology* of films. It is not the viewer and her type of response unavoidably triggered in her mind that are unmistakably philosophical in content but the film itself, *irrespective of the viewer*. In the latter case, there is indeed a venerable variable: Popper's three-world theory of ontology and the corresponding thesis of objective knowledge without a knowing subject. Analogously, *film-philosophy without a viewing subject* suggests that if a nonlinguistic means like a film contains philosophical contents (embodied in it irrespective of its actually being screened), then it may make sense to speak of the possibility of such philosophical contents—abstracta or intelligibles (acknowledged by Jarvie)—as bona fide independent ontological entities that require no mind to entertain them.

Unlike the cognitive reading introduced above, the ontological reading will locate films' direct engagement with philosophy—doing philosophy in an original way independently of linguistic paraphrases—neither in the human cognitive response nor in a mechanical medium unique to the cinema as an art form. Instead, it has to do with a philosophical content somehow embodied in films. We must abstract both from cognition and from the special power of the filmic medium to express philosophical ideas. (The special type of films often used as examples of cinema's ability to do philosophy is called *self-reflexive*, *structural avant-garde* or *minimalist* where the makeup of the film itself is doing the philosophy.) Disregarding the latter approach affords us a view of film as philosophy irrespective either of language as inescapably the privileged medium for doing philosophy or of whether cinematic means must be of a certain kind in order for a film to express genuine philosophical theses.

Looking at what could be specified as the content of the film (to be explained along the Popperian-Bolzano-Fregean line) could open up a way to approach the issue of film as philosophy in a manner that avoids tying down this art form to either the viewing subject (pure cognitivism) or the special properties of the mechanical medium itself (a rather extreme and implausible thesis). If this route is preferred, then Popper's three-world ontology can furnish us the appropriate tools. In addition, the three-world move points to the right direction with regard to one of the poles we wish to avoid, the subject of cognition and the mental acts thereof, given Popper's scheme that subtracts the knowing subject. This is a promising way for getting rid of psychologism intruding into our attempt at establishing the philosophical credentials of movies (though at the end I will defend a kind of cognitivism devoid of psychologism compatible with the World Three view). Anti-psychologism is therefore what we need to adopt.

Anti-psychologism has a ramified history going back to the German philosophical scene of the nineteenth century (see Notturno 1985). At its core, it vehemently shuns psychology from the foundations of mathematics and logic or from the growth of knowledge by means of the sciences. It is a gross mistake to reduce the truths (or laws) of the area we wish to analyze to empirical truths about the psychological inner life of the subject of cognition. As Frege put it in *The Foundations of Arithmetic* (1978, p. 3), it is one thing to know or come to grasp a proposition (that may be a matter of psychology) but quite another—and irreducible to the former—to give a rigorous proof or justification of this proposition.

Jarvie on Films

Jarvie was initially weary of an outright identification of film with philosophy or of films as capable of philosophizing any more than films taken as doing sociology, history, or psychology for that matter. In a relatively early critical survey (Jarvie 1978), he makes clear that movies are not in any sense windows on the world; films are a very poor substitute for arguments or historical or any other inquiry; they cannot even start emulating discursive narratives or historical documentaries. Language, by contrast, can posit alternatives and express objections; complex alternatives can be formulated and argued concisely in linguistic discourses but never by means of films.

However, in his subsequent, book-length, philosophical treatment of film, Jarvie embraces right from the start the thesis that films have contents construed as abstract objects or meanings, and he then goes on to acknowledge that although philosophy is a cognitive activity carried out in a discursive mode while films are complexes of image and sound, it is nevertheless appropriate to say that "non-linguistic sounds and images also contain a good deal of information, including philosophical information" (Jarvie 1987a, pp. 7–8). Given that a philosophical analysis of a film must rest on a linguistic paraphrase to begin with, it is unavoidable that such paraphrasing can never be neutral, and therefore no subsequent interpretation built onto it may be said to be privileged. On the contrary, conflicting or varying interpretations fight it out in a kind of evolutionary struggle (à la Popper's evolutionary epistemology) this being a healthy state of things, a sort of progress. Popper holds that knowledge can grow by competition, eliminating falsified conjectures (as he literally puts it, reminiscent of Nietzsche, Popper 1979, p. 66). Similarly for Jarvie, no interpretation exhausts the philosophical content of the film, and therefore all inherent ambiguity that works of imagination trade in is a sign of strength to be welcomed (Jarvie 1987a, pp. 14–15).

A similar evolutionary approach is further found in Jarvie's piece on the objectivity of criticism in the arts where he sketches what I have called an *institutional* model of objective knowledge-cum-criticism (Jarvie 1987b). Dispensing with all kinds of subjectivism in criticism, objectivity in art criticism is located not in individuals or in statements (texts) but rather in the institutional setting of the practice of criticism that is itself embedded in wider traditions (Jarvie 1987a, p. 312). This allows him to speak of a "democracy of opinion and criticism" (Ibid., p. 214) whereby different viewpoints and evaluations, far from being fruitlessly unbiased (impossible at the individual level), are indeed vying against each other, in a critical rational fashion, within institutional frameworks. The evolutionary approach is evident here (Ibid., p. 184), as opposite theories use critical rational arguments not as justificatory proofs but in order to eliminate rival candidates.

The anti-subjectivist, anti-psychologist viewpoint also informs Jarvie's views on the rationality of creativity and his endorsement of a Popperian World Three model, reiterating the institutional setup necessary for encouraging critical rational thinking conducive to creativity. Creativity is not understood as a psychological mystery or stroke of genius but, in an orthodox Popperian fashion, as involving rational thinking, as a sort of acting and not just as forming beliefs. If creativity has nothing to do with psychology but is a rational, critical, logical activity located in public spaces, the latter is clearly identified with the World Three of objective mind (Jarvie 1985, p. 121). Enlarging the picture to encompass art itself and not just criticism, Jarvie inserts art (along with criticism or theories) into a network of objective relationships whereby the same agonistic fencing of views and theories against rival ones in a rational-critical fashion ensures an evolutionary outcome, in the course of solving problems "constantly subject to world three debate" (Ibid., p. 122). Only within lattices of signification fixed in an objective sense and not as subjective contents of individual minds do theories or art objects like films acquire an identity. "Creativity in a certain sense only exists as a vector in world 3" (Ibid.; Jarvie 1987a, p. 177).

In the course of his book on the philosophy of the film, Jarvie seems to somewhat shift positions compared to the more rigid view adumbrated above: he holds that there *can* be room for films actually doing philosophy. He now admits that films are rather apt in performing "thought-experiments on particular solutions to philosophical problems" in the same evolutionary vein where rival interpretations or "solutions" to such mental experiments vie against each other until one of them seems more rational to uphold. But on those occasions when films do in fact posit radical challenges without any solution to the thought experiment being put forward, he allows that "we have … powerful works of philosophy on film—powerful not so much dramatically, but rather in *translating philosophical dialectic into film*" (Jarvie 1987a, p. 251, emphasis added).

Jarvie's officially stated position is undoubtedly congruent with my interpretation of films qua philosophy as World Three entities. Echoing Popper's view that unasserted or unthought propositions are entities capable of truth or falsity and that World Three entities hide their possible causal import or are potentially dynamically growing or revealing unintended consequence, Jarvie writes: "Roughly speaking, what is philosophically interesting is the revelation of an unnoticed consequence of a philosophy. The production of a new criticism, the detection of inconsistency within a philosophy, or between it and another philosophy not hitherto seen as contradictory, and, perhaps, *a new way of formulating a philosophy*" (Ibid., p.15, emphasis added). This statement must of course be coupled with the following pronouncement which at first sight seems to contradict what has just been quoted for it seems to bring forth the idea of a paraphrase—external to the film itself—as a sine qua non for any philosophical analysis of films to be tenable:

> The other [purpose], which will count with philosophers, is to penetrate the ideas being put forward in items of popular entertainment, to endeavour to state them and subject them to the test of truth. To do this they have to be paraphrased out of their originating context—...—into a context of assessment. We have to be able to identify and label the philosophical components, isolate the arguments implicit in their incorporation in the film, then bring to bear philosophical skill and expertise in assessing these ideas and their argumentative support. When we do this we may come up with somewhat different conclusions than the film offers, *but our tribute to the film consists in its having forced us to think things through for ourselves.* (Ibid., pp. 23–24, emphasis added)

Carefully read, however, I do not think that the above passage forces us to withdraw the general claim that philosophy can be embedded in films. That expertise, philosophical acumen, or other abilities may be required to extract the philosophically relevant content in no way contradicts the thesis of anti-psychologism or the one about abstract contents embodied in filmic medium; after all, the same abilities are required to decipher a text in philosophy too. I now come to an elaboration of a World Three view on films.

Popper via Bolzano and Frege

Popper aimed at putting an end to the false subjectivist theory of knowledge, whereby a subject's beliefs representing the outside world count as knowledge to the extent that they are confirmed. He instead campaigned forcefully for an eternal fallibilism as the kernel of critical rationalism, whereby there can never be complete understanding or knowledge (Popper and Eccles 1977, p. 37). In several places, Popper has propounded a three-world model or tripartite division of entities: a World One of material things and physical processes, a World Two of mental events, psychological attitudes and the like, and a World Three of all the products of the human mind (including scientific theories, customs, mathematics, artworks, airplanes, music, etc.) containing a kind of objective knowledge. Contrasted with subjective knowledge of organisms, objective knowledge concerns knowledge of the logical content of theories (whether published in books, stored in libraries, or discussed), general *objective* problem-solving situations, and human attempts to solve them (failures included and given pride of place in this fallibilist universe) (Popper 1979, p. 73).

I shall come back to Popper's thesis in some detail so as to offer my own distinct proposal with respect to the topic of this chapter, but first I shall briefly dwell on two important predecessors of this way of thinking whom Popper himself mentions (with some important differences among them registered), namely, Bolzano and Frege. I am turning my attention to these two thinkers since what they say, especially about thoughts or propositions as such and the implied anti-psychologism, is more perspicuous than what Popper says on these matters. Some of Popper's inconsistencies and careless formulations need not detain us when filtered through the other two thinkers. At the same time, looking at Bolzano and Frege allows us to follow a more promising route to the desired outcome regarding the plausibility of films being engaged in philosophizing.

Bolzano is mentioned several times by Popper when unfolding his World Three model. *Propositions in themselves*, Bolzano's main protagonists, are mentioned as especially akin to Popper's objective knowledge, and I would add *ideas as such*. Bolzano's *Wissenschaftslehre* (1973) contains one of the earliest and emblematic nineteenth-century theories of propositions (and other semantic items) being divorced from the act of judgment (or of any other subjective mental act) and thus as self-subsisting carriers of meaning. At the same time, this position is crucial for it unequivocally disentangles the content of something (proposition) from its medium (linguistic or otherwise). This is crucial for our purposes as we are dealing with the input a special medium (cinematic) may be able to inject into philosophical knowledge. Bolzano distinguishes a proposition (as such) from the thought of it, from whether it is asserted or whether it may be true or false or it has been put into words (Ibid., §19). Requiring no subject, a proposition as such cannot be said to exist *actually*. More important for present purposes (not mentioned by Popper) is Bolzano's thesis that *ideas in themselves* or *objective* ideas are also removed from actuality; they are never items entertained by a subject during a process of thinking: objective ideas nevertheless exist, as propositions in themselves do, without ever having to be thought of (Ibid., §48). With Bolzano we get one of the earliest statements of what we wish to highlight here: World Three items persist irrespective of any human beings apprehending them; they have semantic force (Bolzano) or constitute objective knowledge; they are like "objects" (Popper).

Bolzano underlines the importance of such ideas for what he calls their "denotative character" (this term designating "reality" as he explicitly claims) and their being absolutely necessary for logic (Ibid., §50). So, abstracta of the kind to be found in World Three are real, and in the case of Bolzano, we also get a semantic-logical route toward their indispensability. (As, similarly, Frege would demand in his lifelong project that proper names must denote, and hence there should be objects designated by them.) The further thesis of Bolzano's that there should be a distinction between words and propositions or between material signs (or neuronal equivalents we would say nowadays in the case of thinking processes) and their abstract denotata is something we need not follow in our case for the medium (art form) is important as well as its cognitive counterpart as we shall see below. This does not jeopardize the thesis of taking films as abstract objects as we shall see presently when we turn to Frege.

Bolzano offers in addition something very useful especially when, as in films and similar art media, we deal with fictional discourse. Given that ideas refer to objects in general and we distinguish an idea from its object, we must be careful when it comes to ideas that refer to nonexisting objects. These are of two types: ideas referring to propositions which are nonexistent (in the sense of non-actual) items or ideas referring to nonexistent objects. Examples of the latter are fictional characters in books, films, or imaginary ones or ones from arithmetic, for example, $\sqrt{-1}$, even ones from areas such as aesthetics or morals, for example, "the highest moral law" (Bolzano) or the "will of the people" as an analogous example used by Frege in *Sense and Reference* (1980, p. 70). In all such cases, the objective idea or idea as such is no less real just because its object does not exist. A similar point is made by Frege in his *Arithmetic* (1978, p. 60).

Despite their differences, Frege is mentioned by Popper as an ally. The seminal paper on *Thoughts* (part I of Frege's *Logical Investigations* 1984 p. 351ff) contains Frege's three-realm model analogous to Popper's, whereby Thoughts are contained in the so-called third realm, being ideas without owners, neither metal states nor external things that can be acted upon as when we grasp them and recognize them to be true. Frege, being a true Platonist (according to standard interpretations followed here), does not hesitate to abolish all change from his third realm and construe its items—Thoughts—as eternal, not human-made, but being discovered, not unlike new planets. This paper of Frege's has been studied extensively, so I propose to take its main anti-psychologistic tenets as more or less well known (establishing the possibility of objective science) and only point out here the following: thoughts can act upon us when they are being grasped without necessarily being true (actually)—namely what is the case with films. So what follows in this section is an elaboration on Frege that is pertinent for fictional (filmic) discourse.

Anti-psychologism and the third realm in Frege's work are to be found not only in *Thought* (1918) but also in his *Logic* of 1897 (Frege 1979, pp. 146–149). Laws of truth as laws of logic are not laws of thought, that is, psychological laws. Thoughts, unlike ideas, are not owned by individual minds; they are nonspatial (Frege 1978, p. 72) and essentially timeless, they are not the products of thinking but only grasped by it: thoughts and laws of logic have nothing to do with psychology. Both logical and mathematical truths, even historical facts, do not depend for their truth (or falsity) on the structure of someone's brain, not "on the chemical composition of his brain" (Ibid., p. 148). The latter point is almost identical to what Jarvie (following Popper) says about a misleading psychologistic analysis of Einstein's mind in order to extract the hidden psychopathology allegedly causally responsible for his unique mental powers and creativity. For Jarvie, the products of such creativity possess their immense significance as items of a World Three cluster of theories and of the objective knowledge lodged in them. Similarly, Einstein's quip that his pencil is more intelligent than him must be understood in terms of the significance and novelty when making certain marks on paper, since both significance and novelty depend in reality not on the physics of his brain pushing the pencil but on the pencil, and its intellectual products, being placed within World Three clusters of scientific theories and the creative rationality characterizing them.

In several places (*Sources of Knowledge of Mathematics and Natural Sciences* 1979, p. 269), Frege highlights an important point related to Bolzano's point about the distinction between content and the bearer of that content that is crucial for our purposes. We distinguish a Thought from a sentence, but the fact we are bound to have Thoughts connected to sentences (a linguistic medium) when we think of them, though inescapable, is nevertheless not a necessary feature of Thoughts, but a feature of our humanity. Thus, it is not that World Three objects are best expressed and formulated in language (as Popper rightly points out) so as to be objects of rational criticism in the course of advancing knowledge, but it just so happens for some of them, primarily scientific theories. It does not mean that other media, like film, cannot be carriers of propositional content or of

Thoughts (Cf. Frege 1984, p. 354; Bolzano makes the same point about language being indispensable for us, but not necessarily for God or other creatures).

A crucial distinction is in order, especially given the sheer Platonism expressed by Frege (despite different views in the relevant scholarship, I here adopt Burge's reading of Frege as sheer Platonist (Burge 1992)). Verifying a proposition about particular things may indeed require us to make observations by means of the senses, but "that proves nothing as to its content" (1978 p. 61). This has to do with relationships between (objective) *concepts*, not with the subjective facts of whether people believe the proposition or not. This is crucial for fictional discourse and for the contents of films (see end of this chapter). A process of abstraction that leads to the formation of a concept precedes the question of whether any number of the items falling under the concept may actually turn out to be nil as in certain fictional characters that are also imaginable ones—they do not exist but only in the make-believe world of a fiction or a film. The objectivity, however, of the concept remains. This Fregean point allows for the possibility of *objective* relationships subsisting between concepts (irrespective of the subjective feelings or thoughts of those entertaining them), even if these are empty of referential content. This is true especially when it comes to deciphering films and their possible philosophical content, even after a Popperian-like eventuality of total destruction of World One or World Two objects (see below). The point mentioned above about the example of the "will of the people" being a term with no generally accepted meaning causing problems of denotation and thus a problem to be avoided in science proper is—alas!—not the case in fictional discourse or filmic content. The opposite is the case. A filmic content needs to be deciphered not despite but because of the inherent ambiguity of terms and what they refer to. In the same work (Frege 1980, p. 63), while talking of truth, Frege remarks that it does not matter whether the name Odysseus has a referent as long as we understand that the poem in which it occurs is a work of art, a fiction (unlike scientific texts). In a footnote on this page, he proposes a special type of signs/terms that have only sense and no referent naming them "representations," and these, he says, are what actors' words on stage are (we could add the same of actors in movies, though not all spoken words by actors are strictly non-referential).

Film as World Three Content

Given the above, I now come to offer a novel reading of the significance of Popper's World Three from a different angle, one that prioritizes the autonomy of this kind of world as well as two of its crucial features. First, according to Popper, a vital characteristic of his World Three of objective knowledge is that it is autonomous from the knowing subject but also autonomous in the sense that new or unintended problems or novel problem situations can emerge and be out of control of the knowing subject (Popper 1979, p. 147). This possibility of constant growth and subsequent impact on us vastly exceeds what we can do to influence it. Here we can place the impact of various art forms, films in particular, as a way of understanding how they (films) can have an impact upon us in completely novel and unforeseen ways. Arguing for the philosophical content of films need not be carried out by conventional means. A film does not need a paraphrase in language for it to contain philosophical theses. Second, Popper argues that self-consciousness or our conscious subjective knowledge depends on theories populating World Three, not the other way round (Ibid., p. 74). World Three is presupposed by World One. Third, Popper claims that knowing involves handling items of objective knowledge (as structures of World Three) analogous to how a builder uses tools to solve an architectural problem or when a bird builds a nest that involves interacting with objective properties of the outside world (Ibid., pp. 113–4, 145, 165–6). This involves coming into contact with objective properties of things—we call them technological affordances nowadays—that is, coming to terms with objective structures and theories of World Three is presupposed. Here we can bring in the Fregean point made above regarding objective relationships of concepts. Being *intelligibles*, World Three items are possible contents of our thought and can be grasped by us.

The second and third points just developed can be employed in tandem to explain how films, their meaning(s), their philosophical import, or their alternative forms of philosophizing (unrolling various subjective stances toward them) must pass through the handling of *intelligibles*. The upshot of all this is that understanding films as World Three items leads us to acknowledge that the structure of films (their special cinematic medium) is itself a special, novel, and unexpected World Three contribution to how we ascribe meanings or read philosophical theses embedded in cinematic modes. Added to this is a further point, congruent with Popper's anti-expressivism in art and epistemology, namely that art is not to be construed as expressing the artists' internal feelings nor is epistemology to be seen as consisting of utterances of subjective beliefs (Ibid., p. 147; Popper 2002, §13).

Popper's three-world ontology revolutionizes Frege's similar division into realms. Where Frege succeeds in showing that no logical or mathematic concepts/thoughts must be reduced to mental images, Popper offers an all-out anti-subjectivist campaign against a suspect epistemological model for the growth of knowledge based on subjective ideas and a knowing subject who is, supposedly, there to justify successive advances in scientific theories and subsequently knowledge in general.

It is true that only certain kinds of filmic means and a certain (limited) extent of philosophy are contained in films. We say "contained," as Jarvie says, and not that films themselves or celluloids or CDs philosophize; human beings do, just as they do with texts: it is not texts that attack each other or disagree on philosophical points. To say that it is contained or that philosophy puts itself in films needs some unpacking: the three-world model shows how to understand this phrase without too much of residual metaphor left behind. It has to do with admitting abstract contents. An added advantage of adopting such a model is that the filmic medium (and certain special techniques employed in the cinema) is thereby seen as one of Popper's objective structures or tools with which to advance knowledge. They are objective structures with structural properties and affordances (like nests) that in no way depend on or vary with subjective feelings or other mental events floating in the minds of film directors or viewers. It is the autonomy of World Three that is important for us here.

World Three items, therefore, are real objects that as thought contents may come into many forms or be embodied in different media. Popper correctly states that if it is a thought content, the abstract idea we are in search of, the one we can abstract, then the medium is not so important because it can be translated into other media (1979, p. 80). This implies that we should not be afraid of paraphrasing after all. If the content remains the same, why should a film medium not contain a thesis with a philosophical import? If criticism and the evolutionary battle of theories are items in World Three (criticism has been ushered in there as part of the critical-rationalist mode of growing knowledge), then critically decoding a thesis (i.e. an abstract thought, a content/meaning) embodied in films by critics makes sense. The thesis itself, when abstracted from its embodiment, is inseparable from the film. In fact, a cinematic embodiment of a thought content of a certain objective idea, critically discussed by a critic in ordinary language, is itself an objective entity.

Cognitivism and Fictionalism

Along with films being part of World Three, Jarvie also advocates the use of science in viewing films, following Ernst Gombrich's lead. Is this an inconsistency? I believe not, but I need to widen the picture. The use of science in analyzing the experience of films is something that has gained momentum, when Anglo-American film scholars turned their attention to neuroscience, psychology, evolutionary theory, and cognitive studies. The import of these empirical sciences, pertaining to the viewer's perceptual fields, is clearly at odds with a purely metaphysical, anti-psycholgistic, picture dealing with films as abstracta. Yet it need not be so. For one thing, World Three itself contains scientific theories, among a rich variety of other items, as well as the practice of rational criticism. It must therefore be of no surprise if cognitivism, which embraces the use of cognitive science to understand film, is itself part of World Three; and so is the criticism of it.

In addition, espousing anti-introspectionism, contemporary cognitivism can be reconciled with World Three or third realm abstracta: the latter can remain a viable ontological model, while the former pertains to the epistemological underpinnings of what goes on in the minds of spectators. But the crucial point is that cognitivism can be compatible with the metaphysical picture of abstract meanings and intelligibles thanks to contemporary scientism that drives cognitivism in film analyses. In other words, Frege's repugnance of mental entities as simply ideas in people's heads or representations that introspection would (illegitimately) promote to admissible epistemological data as the basis of knowledge (mental pictures in flux that could never play the role of stable moorings) is taken care of. Cognitivists in our days employ the whole baggage of neuroscience to reveal permissible stable causal agents that explain how subjects perceive or form mental contents not as a simple unexplainable whimsical triggering of inner psychological life. Generalizations (perhaps denied by some advocates as law-like), not singular truths of how the mind works, are being offered (Currie 2000, pp. 111–112). Lack of space does not allow me to elaborate on how, in my view, a perceptual analysis of imagining what is screened as well as self-placing oneself in the case of a viewer of *Dogville* (given the peculiarity of the explicit bare black-and-white linear theatricality of this particuar film) significantly affects the interpretation of any philosophical content. In fact, this is all compatible with Popper's thesis of World Three items causally impacting on Worlds One and Two (i.e. a scientific theory of perception being one such World Three item).

Some cognitivists have gone so far as to claim that certain film forms can in fact be occasions for people to extend their self-consciousness. Art (and film) is in a special position to extend our human awareness and even make us conscious of new qualia or further enhance our understanding of familiar ones (Smith 2008, pp. 47–50). Odd as it might sound at first, such an attempt is congruent with the Popperian World Three metaphysics given my alternative reading of that model above: for Popper, items in World Three are presupposed by people coming to terms with how they understand themselves and how self-consciousness is constructed (Popper 1979, p. 74). So embrace both cognitivism *and* World Three quasi-Platonism? Yes. This is how Jarvie's bifocal view of films—employing science on the one hand and acknowledging films' abstract objectivity on the other—can be shown to contain no ostensible friction.

As to fictionalism (or the general area investigating various forms of fictional discourse and their semantics and metaphysics) as well as what is called fictionalism in specific areas (e.g., morals), I have space for certain and yet crucial suggestions that affect the argument in an important way as we shall presently see.

First, the above analysis was based on the revised Popperian thought experiment whereby in the event of a total destruction of Worlds One and Two, it is asked: how is it possible to reconstruct or understand the contents of World Three? Having inflated the latter (by Popper himself) so much that almost every human product or activity is included in it, my revised thought experiment asks us to consider: could an alien and mindful species asked to reconstruct all worlds on the basis of coming into contact with only the contents of World Three distinguish fiction from nonfiction? In the case of watching films with the purpose of extracting some philosophical content, it is imperative to clearly make the distinction. How can this be done if what matters for films to philosophize is to regard them as embodiments of abstract contents or objective knowledge?

We need a theory of fictional discourse. There are several alternatives, including a theory of truth in fiction. I register my disagreement with the view in the philosophy of factionalism whereby fictional meaning and fictional knowledge are the same as their ordinary counterparts—something crucial for my thesis here. The second thing to work out is whether such notions and their corresponding theories function differently in the case of films and their sound-and-image medium. Some of the theories of truth in fiction (e.g., Livingston 1996, p. 154ff) would not work in the case of my revised imaginary experiment. Alien minds would not be able to distinguish fiction from nonfiction (e.g., learning from a documentary) without being able to reconstruct the necessary background information (e.g. about historical places, personalities, or even of the physiology of human bodies in sustaining pain); likewise, they would not be able to discern author's intentions (in the sense of Grice's intention-to-believe semantics).

Second, and here fictionalism links up with cognitivism, audiences are very good and quite coherent in making inferences about what is going on in a film even though what they are able to infer is not explicitly contained in the film itself. Extending this, I would ask what inferential networks of relationships could be ascertained without jeopardizing rational prerequisites projected onto film content. In the case of a fictional content, any inferential network of propositions would fall victim to a parallel I propose to the well-known Frege-Geach problem regarding moral expressivism. If a certain proposition has been assigned a truth-in-fiction value in a film or fiction, when does the (possible cognitive) content of such propositions change from being unasserted to being asserted (when moving from unembedded to embedded contexts)? The answer here is linked to anti-expressivism in the arts mentioned above.

Conclusion

The alternative defense of the thesis that films can do philosophy in a genuine way has motivated my going into an elaboration of the Popper three-world thesis via Bolzano and Frege. The alternative looks at the possibility of philosophizing films while avoiding the usual discussion altogether, for I confess that the detractors who cannot accept the possibility of films contributing original philosophical theses always win at the end of the day. The core idea in my proposal has been the following: construed as World Three entities, films can admit of no necessary subject as the one who entertains the meaning, truth(s) or proposition(s), or abstract contents that films contain both in their fictional part and via their medium. In this sense, Jarvie is right in saying (in the motto at the head of this chapter) that nobody really puts philosophy in films; philosophy puts itself in them. His pronouncement, however, needed elaboration. The enrichment of World Three's ontology offered above (by means of Bolzano and Frege) was meant to serve as a necessary basis for an improved and more sophisticated conception of World Three such that it can support this idea that whatever philosophy is "in" the films, it is there as an integral part irrespective of any interpreter or paraphraser expressing thoughts contained therein. To paraphrase Popper: it is a film-philosophy (objectively in films) without a philosophizing or interpreting subject or viewer.

If there is no need for presupposing a subject (or philosophical critic or an interpreter) for 'something in itself' or an 'idea in itself' or a 'proposition in itself' (à la Bolzano) or for a Thought (à la Frege) because such entities can be said to exist without the minds of subjects or viewers, then we can contemplate the following eventuality as a Popperian test: we imagine a total destruction of the world where only World Three items remain and an alien mindful species comes along trying to decipher what our libraries contain in the form of objective knowledge. Following the model reconstructed above, even before the aliens-future interpreters make their appearance these abstracta must subsist. But there is the additional problem of fictionalism, namely, whether certain contents of these libraries, scientific theories, films, documentaries, or literary texts really refer to anything. When deciphered by alien creatures in the absence of Worlds One and Two, the threat of fictionalism poses no problem, because their (abstracta's) truth or falsehood applies only to asserted propositions (à la Bolzano) or judged ones, not to the abstract contents themselves. But for Frege this is different: Thoughts are true or false. This teaches us that following Frege also down *this* path would be a mistake in our analysis.

The alien interpreters can therefore entertain ideas about films in the philosophical mode without falling prey to the problem of distinguishing, in the referential mode, whether fictional entities depicted in the films refer to an outside reality, or more precisely whether they referred to the world before it was destroyed. This would be totally immaterial for the *philosophical* mode of understanding films. But the same is not true for the case of scientific knowledge, and here Popper is wrong: fictionalism is a real threat as far as the science of material things is concerned. In the case of novels or films, however, it does not matter; but in the case of empirical science, it does. It matters e.g. whether the Eigenvalues (and the numbers thereof) in a quantum-mechanical mode refer to a certain physical matrix prevailing in the world the knowers purported to know, describe, and measure realistically (i.e. referentially) by means of their original World Three scientific theories (the aliens are now trying to decipher).

This leads us to the paradoxical conclusion that in the end Popper's World Three is a better description for the products of art but not as appropriate for scientific knowledge. What is needed for understanding non-referentially, in my revised mental experiment, is on the one hand Bolzano's point about ideas that should be accepted without an existing object but with their material as propositions (the abstract contents of films, let us say) without coupling it on the other hand with Frege's point about the objective relationships of objective (referring) concepts.

A final point stemming from the overarching anti-psychologism inspiring the protagonists behind the analysis offered here: The Popper-Bolzano-Frege anti-psychologism brings to the fore the concomitant distinction between subjective and objective in a quasi-Platonist mode. It is extended and perhaps reformulated in Jarvie's view of institutions-practices-traditions being responsible for distinguishing the subjective thoughts (e.g., of critics) from the objective significance of propositions in books and thoughts expressed in language.

References

Bolzano, Bernard. 1973 [1837]. *Theory of Science*, Trans. B Terrell. Dordrecht: Reidel.

Burge, Tyler. 1992. Frege on Knowing the Third Realm. *Mind* 101 (404): 633–650.

Currie, Gregory. 2000. Cognitivism. In *Companion to Film Theory*, ed. Toby Miller and Robert Stam, 105–122. Oxford: Blackwell.

Frege, Gottlob. 1978 [1884]. *The Foundations of Arithmetic*. Translated by J.L. Austin. Oxford: Blackwell.

———. 1979. *Posthumous Writings*. Edited by Hans Hermes et al. Oxford: Blackwell.

———. 1980. *Translations form the Philosophical Writings of Gottlob Frege*. Edited by Peter Geach et al. Oxford: Blackwell.

———. 1984. *Collected Papers on Mathematics, Logic and Philosophy*. Edited by B. McGuiness. Oxford: Blackwell.

Jarvie, Ian C. 1978. Seeing through Movies. *Philosophy of the Social Sciences* 8 (4): 374–397.

———. 1985. The Rationality of Creativity. In *The Concept of Creativity in Science and Art*, ed. Dennis Dutton and Michael Krausz, 109–128. Dordrecht: Martinus Nijhoff.

———. 1987a. *Philosophy of the Film: Epistemology, Ontology, Aesthetics*. London: Routledge.

———. 1987b. The Objectivity of Criticism of the Arts. In *Rationality: The Critical View*, ed. Joseph Agassi and Ian C. Jarvie, 201–216. Dordrecht: Martinus Nijhoff.

Livingston, Paisley. 1996. Characterization and Fictional Truth in the Cinema. In *Post Theory: Reconstructing Film Studies*, ed. David Bordwell and Noel Carroll, 149–174. Madison: The University of Wisconsin Press.

Notturno, Mark Amadeus. 1985. *Objectivity Rationality and the Third Realm: Justification and the Grounds for Psychologism: A Study of Frege and Popper*. Dordrecht: Martinus Nijhoff.

Popper, Karl. 1979. *Objective Knowledge*. Oxford: Clarendon Press.

———. 2002. *Unended Quest: An Intellectual Autobiography*. London: Routledge.

Popper, Karl, and John Eccles. 1977. *The Self and Its Brain*. Berlin: Springer.

Smith, Murray. 2008. Consciousness. In *The Routledge Companion to Philosophy and Film*, ed. Paisley Livingston and Carl Plantiga, 39–51. London and New York: Routledge.

Wartenberg, Thomas. 2007. *Thinking on Screen: Film as Philosophy*. New York: Routledge.

CHAPTER 25

Jarvie, Popper, McLuhan, and Me

Paul Levinson

I first met Ian C. Jarvie when I was putting together and editing *In Pursuit of Truth: Essays on the Philosophy of Karl Popper on the Occasion of His 80th Birthday* (1982), a Festschrift, and my first and only edited book until *The Essential Sopranos Reader* (2011), indicating my overwhelming preference for being an author rather than an editor of books. Although I don't know if Jarvie ever wrote about "The Sopranos" per se, he did write about television—see his *History on Television* (2001)—and at least half of his life's work has been about the epistemology and business of movies; so, his work in effect provides a bridge between the first and recent collection of essays I edited.

Back in 1980 and 1981, I didn't know what to expect when I was putting together *In Pursuit of Truth.* I was a media theorist, not a philosopher by academic training, though my work, especially my doctoral dissertation, *Human Replay: A Theory of the Evolution of Media* (1979), was beginning to be accepted in philosophy-of-technology circles. Indeed, it was Karl Popper's (1972) "World 3" categorization of ideas and technologies—products of the human mind (World 2) which exist independently of their creators—which first brought me in touch with Popper and his ideas, and in turn Ian C. Jarvie, as I began work on the Popper Festschrift.

P. Levinson (✉)
Department of Communication & Media Studies, Fordham University, Bronx, NY, USA

R. Sassower, N. Laor (eds.), *The Impact of Critical Rationalism*,
https://doi.org/10.1007/978-3-319-90826-7_25

Most of the other contributors to that volume were traditional philosophers, ranging from Gerard Radnitzky to W. W. Bartley III, who had studied, written about, commented upon, and in many cases, expanded Popper's philosophy of science, social philosophy, and epistemology. Jarvie was also a traditional philosopher, with a specialty in social philosophy, but he also had something which I'm pretty sure none of the other contributors to *In Pursuit of Truth* had: an understanding of and conversance with Marshall McLuhan and his unique assessment of media. This accompanied and was no doubt stoked by Jarvie's deep interest in the films of Woody Allen (see Jarvie 2004), the centrality of *Annie Hall* (1977), and McLuhan's immediately iconic appearance in the movie to hilariously denounce the fictitious Columbia University professor who, like so many real professors in the world, utterly misunderstood McLuhan's work.

Jarvie, in contrast, got McLuhan, and that meant a lot to me. In a hotel lobby, shortly after the publication of *In Pursuit of Truth*, I was explaining McLuhan (even though he always insisted "I don't explain—I explore"; see McLuhan 1967) to about half of the contributors to the Popper volume. All of them, except Jarvie, gave me courteous nods. Jarvie jumped into the conversation.

In the years that followed, I wrote more about McLuhan than Popper—though my *Mind at Large: Knowing in the Technological Age* (1988) extensively discussed both—but my interest in media theory and Jarvie's about film brought us together for several conferences, especially the International Conferences on Culture and Communication organized by Sari Thomas at Temple University in Philadelphia in the 1980s, where Jarvie and I could often be found sitting in the back of a packed room, kibitzing about everyone from Hugo Münsterberg to the Beatles.

Philosophy is often mischaracterized as esoteric or irrelevant to the lives we live in the everyday world—or maybe correctly characterized as that for many who are recognized as professional philosophers, whatever exactly that may be. Ian C. Jarvie has always been very different, a twentieth- and now twenty-first century example of the kind of philosopher Socrates likely was on the ancient streets of Athens. I'm glad to see this Festschrift dedicated to Jarvie, and I'm glad to know him.

References

Allen, Woody. 1977. Annie Hall. Movie Distributed by United Artists.

Jarvie, Ian C. 2001. History on Television. *Historical Journal of Film Radio and Television* 21 (1): 97–99.

———. 2004. Arguing Interpretation: The Pragmatic Optimism of Woody Allen. In *Woody Allen and Philosophy: You Mean My Whole Fallacy Is Wrong?* ed. Aeon J. Skoble and Mark T. Conard, 48–65. Chicago: Open Court.

Levinson, Paul. 1979. *Human Replay: A Theory of the Evolution of Media*. PhD dissertation, New York University. Reprinted by *Connected Editions*, 2017.

———, ed. 1982. *In Pursuit of Truth: Essays on the Philosophy of Karl Popper on the Occasion of His 80th Birthday*. Atlantic Highlands, NJ: Humanities Press.

———. 1988. *Mind at Large: Knowing in the Technological Age*. Greenwich, CT: JAI Press.

Levinson, Paul, David Lavery, and Douglas Howards, eds. 2011. *The Essential Sopranos Reader*. Lexington, KY: University of Kentucky Press.

McLuhan, Marshall. 1967. Casting My Perils Before Swains. In *McLuhan: Hot and Cool*, ed. Gerald Stearn. New York: Dial.

Popper, Karl R. 1972. *Objective Knowledge*. Oxford: Oxford University Press.

Index[1]

[1] Note: Page numbers followed by "n" refers to notes.

R. Sassower, N. Laor (eds.), *The Impact of Critical Rationalism*,
https://doi.org/10.1007/978-3-319-90826-7

The manufacturer's authorised representative in the EU is Springer Nature Customer Service Centre GmbH, Europaplatz 3, 69115 Heidelberg, Germany. If you have any concerns regarding our products, please contact ProductSafety@springernature.com

Printed and bound by CPI Group (UK) Ltd, Croydon, CR0 4YY
30/09/2026
02225597-0009